Notes and Letters on the Natural History of Norfolk

by Thomas Browne

I0493038

CONTENTS.

INTRODUCTION.

"Every kingdom, every province, should have its own monographer."-- Gilbert White. Seventh Letter to Barrington.

The excellent Memoir of Sir Thomas Browne, in Wilkin's Edition of his works, renders it unnecessary here to repeat what has already been so well done; suffice it to say that he was born in London on the 19th of October, 1605; he was educated at Winchester School and entered at Broadgates Hall (now Pembroke College), Oxford, in 1623; graduated B.A. 31st January, 1626-7, and M.A. 11th June, 1629. About the year 1633 he was created Doctor of Physick at Leyden. In 1636 he took up his residence in Norwich, in 1637 was incorporated Doctor of Physic in Oxford, and in 1665 was chosen an Honorary Fellow of the College of Physicians. In 1671 Browne was knighted at Norwich by Charles II., and after a useful and honourable career died on his seventy-sixth birthday, the 19th of October, 1682, and his body lies buried in the church of St. Peter Mancroft, Norwich.

Browne in early life travelled much and was a voluminous writer; he made many friendships with men celebrated in his day, and his advice and assistance were sought and gratefully acknowledged by Dugdale, Evelyn, Ray and Willughby, Merrett, Sir Robert Paston (afterwards Earl of Yarmouth), Ashmole, Aubrey, and others; but his general correspondence does not now concern us, my object being to supply in a convenient form what I believe will be acceptable to modern naturalists, namely, an accurate transcript of his notes and letters on the "Natural History of the County of Norfolk."

These notes and letters were first published by Simon Wilkin in his Edition of Sir Thomas Browne's Works in 1835, but they were not treated from a naturalist's point of view, and in some places were not correctly transcribed, added to which, in the vast mass of matter contained in Wilkin's four large volumes (or in the closely printed three volumes of Bohn's Edition), these interesting passages are in danger of being overlooked or are inconvenient for reference. Two letters, moreover, were needed to make the correspondence with Merrett complete, and these I have been enabled to supply. I hope also that my explanatory notes, which I trust will not be deemed too voluminous, will be found more useful than the necessarily brief notes furnished by Wilkin and his collaborators. Furthermore, I think that the retention of the original spelling and punctuation may lend a charm to the quaintness of the language which is in a measure destroyed by any attempt at modernising.

There is much that is interesting bearing upon Natural Science scattered throughout Browne's writings, especially in his Pseudodoxia Epidemica, or inquiries into Vulgar and Common Errors, first published in 1646, and the reader cannot fail to be impressed not only with the extent of his classical knowledge but also with the shrewdness with which he pursued his original investigations; but here it is only proposed to deal with certain manuscript notes and a series of rough notes for, or copies of, letters addressed to Dr. Christopher Merrett, the author of the Pinax Rerum Naturalium Britannicarum. These, as remarked by their editor, with regard to some other manuscripts published[A] in 1684, under the title of "Certain Miscellany Tracts," were doubtless "rather the diversions than the Labours of his Pen; and ... He did, as it were, drop down his Thoughts of a sudden, in those spaces of vacancy which he snatch'd from those very many occasions which gave him hourly interruption;" but I cannot in this instance agree with the conclusion arrived at by the same writer that it "seemeth probable that He designed them for publick use," for they appear to be the rough drafts or memoranda used in the production of the finished letters (which are unfortunately not forthcoming), and were never intended for publication in their present crude form, thus rendering pardonable such annotations as I have ventured to add. But before proceeding further it is necessary to consider briefly the time and circumstances under which they were written, and the state of what passed for Natural Science at that period.

[A] The "Miscellany Tracts" were put forth by "Tho. Tenison" (1636-1715), who afterwards became Archbishop of Canterbury, but was then the Rector of a London parish, St. Martin-in-the-Fields. He had been a Norwich school-boy, and subsequently minister of St. Peter's Mancroft. He was doubtless well acquainted with Browne and his family, and hence his reference in the preface quoted to "the Lady and Son of the excellent Authour," who, he says, "deliver'd" the papers to him.

Browne wrote early in the second half of the seventeenth century, during a period of great awakening in the study of Nature. Hitherto it could hardly be said that a direct appeal to the works of Nature had been the prevailing method. Aristotle was still the established authority, and commentaries on his works occupied the minds of men to the exclusion of original investigation, notwithstanding that this great philosopher had himself, both by precept and example, urged the importance of direct observation and inquiry; the Mediaeval school of thought still prevailed and cramped every effort at progress. How keenly Browne lamented this spirit of slavish adherence to tradition may be judged from a passage in one of his Essays in the "Vulgar Errors" condemning the obstinate adherence unto antiquity; he writes, "but the mortallist enemy unto knowledge, and that which hath done the greatest execution upon truth, hath been a peremptory adhesion unto authority; and more especially the establishing of our belief upon the dictates of antiquity. For (as every capacity may observe) most men of ages present, so supersticiously do look upon ages past, that the authorities of one exceed the reason of the other." In another place he argues that the present should be the age of authority, seeing that we possess all the wisdom of the ancients which has come down to us, with that of our own times added. In fact, Browne's motto appears to have been "prove all things and hold fast only to that which is good."[B]

[B] There was one form of ancient authority before which Browne bowed down with absolute and unquestioning submission--the authority of the Scriptures. In all secular matters he was ever ready to point the lance and do battle, but all that appealed to him on what he regarded as divine authority was beyond the pale, and it never entered into his mind to submit it to the test of reason. In the "Religio Medici" he declares his devoted adherence first to the guidance of Scripture, and secondly to the Articles of the Church, "whatsoever is beyond, as points indifferent, I observe according to the rules of my private reason;" and again, "where the Scripture is silent, the Church is my text; where that speaks 'tis but my comment; where there is a joint silence of both I borrow not the rules of my religion from Rome or Geneva, but the dictates of my own reason." This implicit adherence to the literal text of Scripture led to his--shall I say active belief in, or passive acceptance of, the existence of Witchcraft, and thus to the only act in an otherwise blameless life which we must regard with regret and astonishment. I refer to the consenting part he took in the doing to death of two poor women at Bury St. Edmund's in the year 1664. It is my business to act as Browne's exponent, not as his apologist, but it must be borne in mind that in his day the "higher criticism" was a thing unheard of, and that the literal sense of the English translation of the Bible was accepted as binding not only by him but by the vast majority of the people, including the most learned men of the time. "Thou shalt not suffer a witch to live" was a plain command, and given a witch the believer's duty was also plain; that there had been witches there was ample scriptural evidence, but there was none that the days of witchcraft had passed away. Browne only shared this belief with his pious friend, the venerable Bishop Hall, and many men equally devout according to their lights; he makes no secret of the fact and acts in accordance with his convictions and the plain authority of Scripture. Thus it came about that these conscientious but mistaken men were induced to render possible, if not actually to countenance, the fiendish cruelties perpetrated by their unscrupulous allies. In matters which he considered less authoritative his views were so liberal as to gain for him the stigma of infidel or heretic; but let a man govern his thoughts and actions by the private rules Browne laid down for his own guidance (vol. iv., p. 420), and it would be hard to regard him as otherwise than a God-fearing man, striving to live up to his profession.

Aristotle, whose works on Natural History have descended to us in a very imperfect condition, lived in 385-322 B.C., and it was not till A.D. 79 that the Historia Naturalis of Pliny the Elder the next great work, which has survived till our days, was completed, and by some of those most competent to form a judgment the additions which he made were not in all cases improvements. Other writers followed, but their productions were of little value, and it was not till the year 1544 that William Turner published at Cologne what Professor Newton describes as "the first commentary on the birds mentioned by Aristotle and Pliny conceived in anything like the spirit that moves modern Naturalists." Turner's book is very rare and unfortunately at present beyond the reach of most modern students. No attempt at systematic arrangement, as now understood, was made until the Histoire de la Nature des Oyseaux of Pierre Belon (Bellonius) appeared at Paris in 1555, for the much greater work of Conrad Gesner, being the third book of his Historia Animalium, which was published at Zurich in the same year, and treated of Birds, followed, more or less closely, an alphabetical plan which brought upon him the censure of Aldrovandus, three of whose sixteen folio volumes forming the Historia Naturalium bore the title of Ornithologiae hoc est de Avibus Historiae, Libri XII., and were brought out at Bologna between the years 1599 and 1603. The Historia Naturalis of John Jonston, or "Jonstonus" (1603-1675), originally published in four sections between the years 1649 and 1653, ran through several editions, and was a popular book in the seventeenth century; it is frequently referred to by Browne, but is a work of very little originality.

Though all these authors undoubtedly influenced their successors, it may be fairly said that it was Browne's contemporaries and fellow-countrymen, Francis Willughby and John Ray, who laid the first solid foundation of systematic zoology in their Ornithologia and Historia Piscium, published in 1676 and 1686 respectively; but dying in 1682, Browne was indebted to neither of them, though he doubtless exercised much influence over them, and he had to use the clumsy descriptive terminology then in vogue.[C] Let me illustrate this by a single example. In one of his letters to Merrett he names a "little elegant sea plant" (probably Halecium halecinum, a species of Hydroid Zoophyte), "Fucus marinus vertebratus pisciculi spinum referens ichthyorachius, or what you think fit." On another occasion Merrett thus expresses his approval of Browne's efforts in this direction: "You have very well named the rutilus and expressed fully the cours to bee taken in the imposition of names, viz: the most obvious and most peculiar difference to the ey or any other sens." We can hardly conceive the difficulties these pioneers of Natural Science had to contend with; the works of their predecessors were so indefinite as to be of little value in determining species; they had to depend upon the vague descriptions of fowlers and others; the same bird would probably be known in half a dozen different localities by as many different names, and since no satisfactory mode of preserving specimens had then been discovered, examples for comparison were not available. If inextricable confusion arose with regard to such a bird as the Osprey, well might Browne write with regard to those less readily characterized, "I confess for such little birds I am much unsatisfied on the names given to many by countrymen, and uncertaine what to give them myself, or to what classis of authors cleerly to reduce them. Surely there are many found among us which are not described; and therefore such which you cannot well reduce, may (if at all) be set down after the exacter nomination of small birds as yet of uncertain class of knowledge."

[C] In 1735 appeared the first edition of the Systema Naturae of Linnaeus which, meagre as it was, ushered in a more definite system of classification, whilst his invention of the binomial method of nomenclature, first used by him in the tenth edition of that work published in 1758, contributed not a little in reducing to order what had hitherto been a chaos, although in his classification of birds he for the most part followed his predecessor Ray.

I must ask pardon for this digression, but my object has been to show the difficulties Browne had to contend with and to emphasise the originality which pervades all his observations, a characteristic so conspicuously absent in the work of most of his predecessors. I should like also to call attention to his references to the migratory habits of many species of birds, a phenomenon attracting little notice in his day, but one which can be so readily observed on the coast of Norfolk. These remarks were penned at a time when hibernation in a state of torpidity was thoroughly believed in--an idea of which even Gilbert White a hundred years later could not thoroughly divest himself. In his tract on "Hawks and Falconry," Browne further says: "How far the hawks, merlins, and wild-fowl which come unto us with a north-west [east?] wind in Autumn, fly in a day, there is no clear account: but coming over the sea their flight hath been long or very speedy. For I have known them to light so weary on the coast, that many have been taken with dogs, and some knocked down with staves and stones." Further than this, he knew the seasons of their appearing-- the Hobby "coming to us in the spring," the Merlin "about autumn." His frequent mention of anatomical peculiarities and of his dissections of many birds and beasts clearly prove his passion for original research, and the frequent records of the contents of the stomachs of the birds which he had the opportunity of examining was a mode of obtaining exact information as to the nature of their food, which I imagine was not common in those days.

How highly Browne was esteemed by his contemporaries may be judged from the acknowledgments of his assistance by Dugdale, Evelyn (who visited him in Norwich in 1671), and others; and Ray especially mentions his indebtedness to "the deservedly famous Sir Thomas Browne, Professor of Physic in the City of Norwich." His letters to his son, Dr. Edward Browne, are full of instructions as to the course of study he should pursue, and subsequently, when the latter became celebrated and was appointed Physician to St. Bartholomew's Hospital, it was still to his father that he looked for advice in his hospital practice and in the preparation of his lectures. Browne was proud of his adopted county, a feeling evidently shared by his son, and I trust I may be pardoned for quoting the concluding passage of the latter's account of a tour into Derbyshire, wherein he expresses a sentiment which survives with undiminished force in the breast of many a Norfolk man in the present day. There is a very interesting account of his crossing the Wash on leaving Lynn for Boston, but on his return to Norwich in September, 1662, he thus concludes his journal: "Give me leave to say this much: let any stranger find mee out so pleasant a country, such good way [roads], large heath, three such places as Norwich. Yar [Yarmouth] and Lin [Lynn], in any county of England, and I'll bee once again a vagabond to visit them."

The manuscripts of which the following selection forms a part are contained, with a few exceptions to be named hereafter, in the Sloane Collection in the Library of the British Museum, consisting of nearly one hundred volumes, numbered 1825 to 1923 both inclusive. A catalogue is given by Simon Wilkin[D] (himself a Norfolk man), by whom Browne's collected writings were first published in a connected form, as already mentioned, under the title of "Sir Thomas Browne's Works, including his Life and Correspondence, edited by Simon Wilkin, F.L.S. London, William Pickering. Josiah Fletcher, Norwich, 1836." 4 volumes, 8vo; the first volume only is dated 1836, Vols. 2, 3, and 4 being dated 1835.[E] It was here that the Notes and Letters were first given to the public. A second edition of the "Works," also edited by Wilkin, in three closely printed volumes, was issued in Bohn's Antiquarian Library in 1852. In the first edition the Notes on the Birds and Fishes will be found in Vol. IV., pp. 313 to 336, and the letters to Merrett in Vol. I., pp. 393 to 408. In the second edition both are in Vol. III., pp. 311 to 335 and pp. 502 to 513 respectively. The references here, as a rule, will be made to the 1836 edition, when otherwise Bohn's edition will be specified.

[D] Simon Wilkin (1790-1862), the able editor of Sir Thomas Browne's collected works, was born at Costessey near Norwich, in the year 1790. He came to Norwich after his father's death in 1799, taking up his temporary abode with his guardian, Joseph Kinghorn, a Baptist minister of note and a prominent member of a literary circle then existing in Norwich, by whom his education was superintended. On arriving at man's estate and being at that time possessed of ample means, he devoted himself to the study of Natural History, especially to Entomology, and was the possessor of a large collection of insects which, in the year 1827, was purchased for the Norwich Museum at a cost of one hundred guineas, a large sum in those days. He was one of the founders and the first librarian of the Norfolk and Norwich Literary Institution in 1822, also of the Norfolk and Norwich Museum in 1825, both of which institutions (the former reunited to its parent Library, founded in 1784) are still flourishing. Wilkin was a Fellow of the Linnean Society, also a Member of the Wernerian Society of Edinburgh. In later years the loss of the bulk of his property by a commercial failure necessitated his turning his attention to some means of earning a livelihood, and he established himself in Norwich as a printer and publisher; later in life he removed to Hampstead, where he died on 28th July, 1862, and was buried in his native village of Costessey.

[E] Some copies of this Edition have a title-page, bearing the name of H. G. Bohn as publisher, and the date of 1846, but differing only in that respect.

The foot-notes in Wilkin's edition, many of them very curious, initialled "Wr.," are by Dr. Christopher Wren, Dean of Windsor (father of the Architect of St. Paul's Cathedral), and were found on the margins of a copy of the first edition of the Pseudodoxia now preserved in the Bodleian Library at Oxford; those initialled "G." were written for Wilkin's first edition by the late Miss Anna Gurney, of Northrepps, near Cromer, Norfolk.

The first papers to which I shall refer are a series of rough notes contained for the most part in volume 1830 of the Sloane MSS., the first portion being devoted to Birds found in Norfolk, followed by a similar series relating to marine and freshwater Fishes, including a few marine invertebrata and plants. They are written on one side only of foolscap paper, the portion relating to Birds occupying folios 5 to 19 inclusive, folios 1 to 4 consist of two inserted letters from Merrett to Browne (see Appendix A.), which are printed by Wilkin in his first edition, Vol. I., pp. 442-5. The notes on Fishes are in the same volume of manuscripts, folios 23 to 38; but there are some irregularities which will be explained as they occur. The whole of the notes are very roughly written, and present the appearance of a commonplace book, in which the entries were made as the events occurred to the writer, being quite devoid of any system or arrangement. The entries doubtless extend over several years, but it is impossible to fix the dates on which they were made, the only internal evidence I can find being that speaking of the occurrence of a certain shark he states it was taken "this year, 1662," and on the next page of the MS. there is the record of the occurrence of a sun-fish in the year 1667; this latter, however, is evidently an interpolation. A few pages further on there is the record of what he calls a large mackerel, "taken this year, 1668," but this also is an addition. We may take it, I think, that most of the notes were made about the year 1662, but that they were added to on various occasions up to 1668, in which year his first letter to Merrett is dated. It has been suggested that these notes were prepared in the interest of Dr. Merrett for his use in an enlarged edition of his Pinax, but the remark in his first letter to this correspondent, "I have observed and taken notice of many animals in these parts whereof 3 years agoe a learned gentleman of this country wished me to give him some account, which while I was doing ye gentleman my good friend died," clearly shows that they were originally prepared for another purpose, although they eventually furnished the materials for his letters to Merrett, but who his deceased friend was it seems now useless to conjecture, although it would be interesting to know. The notes were certainly never intended to appear in their present form, and failing their use by Merrett which never took place, the information they contained was, as we know, of great service to Ray and Willughby.

Browne's correspondent, Dr. Christopher Merrett, was born at Winchcomb, in Gloucestershire, on the 16th of February, 1614. He graduated B.A. at Oriel College, Oxford, about the year 1635; M.B. 1636; M.D. 1643. Was elected Fellow of the Royal College of Physicians in 1651, and was made first Keeper of the Library and Museum; he was Censor of the College seven times. Having entered into litigation with the College with regard to his appointment, which was considered by that body to have terminated when the Library was destroyed by the great fire, he was defeated, and in 1681 expelled from his fellowship. He died in London in 1695. ("Dict. of Nat. Biog.") Merrett was the author of several works on various subjects, as well as of the Pinax, and a translation of the "Art of Glass" referred to further on. His Pinax Rerum Naturalium Britannicarum, said to have been brought out in 1666, contained the earliest list of British Birds ever published, but it is little more than a bare list. Copies bearing the date of 1666 are very rare, and it is believed the edition was burned in a fire at the publishers; but Professor Newton ("Dict. of Birds," Introduction, p. xviii.) says that in 1667 there were two issues of a reprint; one, nominally a second edition, only differs from the others in having a new title-page, an example doubtless of what Wilkin severely condemns as "that contemptible form of lying under which publishers have endeavoured to persuade the public of the rapidity of their sales." Merrett was contemplating a new and improved edition of his work when, as Wilkin happily puts it, "in an auspicious moment he sought the assistance of Browne, whose liberal response is evidenced in the [drafts of the] letters still fortunately extant, but either superseded by the more learned labours of Willughby and Ray, or laid aside on account of the perplexities in which Merrett became involved with the College of Physicians, the Pinax never attained an enlarged edition. Had Browne completed and published his own 'Natural History of Norfolk,' he might have contended for precedency among the writers of County Natural Histories with [his friend] Dr. Robert Plot,[F] who published the earliest of such works--those of Oxford and Staffordshire, in 1677 and 1686 respectively. He seems, however, to have preferred contributing to the labours of those whom he considered better naturalists than himself; and in his third attempt thus to render his observations useful he had somewhat better success. He placed his materials, including a number of coloured drawings, at the disposal of Ray, the father of systematic Natural History in Great Britain, who has acknowledged the assistance he derived from him in his editions of Willughby's 'Ornithology' and 'Ichthyology,' especially in the former. But Browne, it seems, found it more easy to lend than to recover such materials; for he complains, several years afterwards, that these drawings, of whose safe return he was assured, both by Ray and by their mutual friend, Sir Philip Skippon, had not been sent back to him."[G]

[F] Dr. Robert Plot (1640-1696) was born at Sutton Barne, Kent, in 1640; he graduated M.A. in 1664, and D.C.L. at Oxford in 1671. He was chiefly noted as an antiquary, and was Secretary of the Royal Society from 1682 to 1684, also the first custodian of the Ashmolean Museum and Professor of Chemistry at Oxford. In 1677 he published his "Natural History of Oxfordshire," the first local work of the kind which appeared; it was illustrated by sixteen plates. In 1686 he also published "The Natural History of Staffordshire," and subsequently many other books and papers. He was evidently acquainted with most of the learned men of his time. Plot died at his family estate Sutton Barne, on the 30th of April, 1696, and was buried at Borden in Kent. Dr. Plot was a friend of Browne's, and his companion in a tour in England in 1693.--"Dict. Nat. Biog."

[G] See letter to his son, Dr. Edward Browne (Wilkin, i., p. 337), also Appendix C.

I have endeavoured to reproduce as accurately as possible the text of the notes and letters, which, as will be seen from the example photographed for the frontispiece of this volume, was often very difficult to decipher. The originals of the notes and of seven of the nine letters to Merrett, as also the two letters in Appendix A., are in the Sloane Collection of MSS. in the British Museum Library; those numbered vii. and viii., as well as two letters in Appendix D., which have not hitherto been printed, are in the Bodleian Library; and the letter to Dugdale in Appendix B. is extracted from the "Eastern Counties Collectanea." All the MSS. in the Sloane Collection I have transcribed myself; of those in the Bodleian Library, No. vii. is from a photograph, the remainder were copied for me by a person recommended as being highly reliable. I thought it best to retain all the erasures and interlineations in order to show as much as possible what was passing in their author's mind: in the foot-notes I have sought to acknowledge in situ the valuable help I received from numerous correspondents to whom my best thanks are due, but I owe a special debt of gratitude to Professor Newton, at whose instigation the work was undertaken, for his kind assistance and for the loan of scarce books which it was necessary to consult in the interesting investigations needful to elucidate, if possible, some of the obscure passages in the text, a task in which if with the best intentions should I have sometimes failed, I must ask the reader's indulgence.

It may be truly said of Sir Thomas Browne that a prophet hath no honour in his own country; the writings of this remarkable man are little known in the city of his adoption, and a recent movement to erect a monument to his memory has hitherto met with feeble support.

T. S.

Norwich, December, 1901.

Notes and Letters

ON THE

Natural History of Norfolk.

NOTES[H] ON CERTAIN BIRDS FOUND IN NORFOLK.

[H] The heading adopted by Wilkin, for which I cannot find that he had any authority, is certainly misleading, for the brief and fragmentary notes which follow, although of great interest, can hardly be called "An Account of the Birds (or Fishes) found in Norfolk," as there are many species of each inhabiting or visiting the county which must have been well-known to Browne, but of which we find no mention.

[MSS. SLOAN. 1830. FOL. 5-19. AND 31.]

[The first four pages in the volume of Manuscript consist of two inserted letters from Merrett to Browne (see Appendix A.); these are on ordinary letter paper 6-1/4 inches by 7 inches. The notes commence on folio 5 and are continued to folio 19; one leaf, containing an account of the Roller (numbered 31), is bound up with the notes on the Fishes, &c., which are numbered consecutively with the Birds; the paper of the volume is foolscap, 11-1/2 by 7-1/2 inches, and written, with a few exceptions, which appear to be subsequent additions, on the right-hand opening only. There are four folios after the Birds, the first of which is blank; the others, numbered 20, 21, and 22, contain rough memoranda on the Birds and Fishes, the substance of which is embodied in the other notes; the Fishes commence on folio 23. There are many erasures, interlineations, and substituted words which indicate hasty writing, and the alterations are not in all cases complete, thus rendering the sense occasionally obscure; these emendations I have thought it best to preserve as indicating the author's line of thought. In the foot-notes which follow I have endeavoured to identify the species treated of. This, notwithstanding the kind assistance of the friends whose help I gratefully acknowledge, I may not in all cases have successfully accomplished; the conclusions arrived at are occasionally only conjectural, and it may be that in some instances I have erred. Should such be the case I must plead in excuse the difficulty arising from vagueness of description, the frequent use of vernacular names which have long since become obsolete (see Note 22), and the imperfection of the record. This especially applies to the Marine Animals, and one of my correspondents rightly remarks that "the early accounts of marine beasts are so vague, and the figures (where referred to) so incomplete and often fanciful, that it is difficult even to make out the family, to say nothing of genera and species." Any assistance or correction in this respect would be gladly received by me.]

[Fol. 5.] I willingly obey your comands[1] in setting down such birds fishes & other animals wch for many years I have observed in Norfolk.

[1] With regard to the probable origin of these notes (see "Introduction," p. xxi.). The opening passage was probably addressed to the deceased correspondent who had asked his assistance, whereas his first letter to Merrett seems to indicate that the offer of help to him came spontaneously from Browne ("I take ye boldness to salute you," &c.), and was not in response to Merrett's request.

Beside the ordinarie birds which keep constantly in the country many are discouerable both in winter & summer wch are of a migrant nature & exchange their seats according to the season.[2] those wch come in the spring coming for the most part from the southward those wch come in the Autumn or winter from the northward. so that they are obserued to come in great flocks with a north east wind & to depart with a south west. nor to come [in struck out] only in flocks of one kind butt teals woodcocks felfars thrushes & small birds to come & light together. for the most part some hawkes & birds of pray attending them.

[2] Browne seems to have had on the whole a fairly correct idea with regard to the migratory movements of the birds on the Norfolk coast where peculiar facilities exist for such observations, but of course he could have formed no notion of the extent to which they prevail, perhaps no species being altogether sedentary. The general line of the autumn migration for those which spend their summer in Northern Europe is south or south-west, returning in the spring by the reverse route; those which visit us in spring from Western Europe, or countries lying still more to the eastward, adopt what is known as the east to west route, and reverse the direction in the autumn; but this latter is as nothing compared with the vast number of immigrants by both routes in the early autumn, at which time, especially, the movements are so exceedingly complex that it would be impossible here to attempt to explain them, and the reader must be referred to Mr. Eagle Clarke's digest of the Reports of the Migration Committee of the British Association ("Report Brit. Ass. for 1876," pp. 451-477).

The great & noble kind of Agle calld Aquila Gesneri[3] I have not seen in this country but one I met with [with crossed out] in this country brought from ireland wch I [presented unto struck out] kept 2 yeares, feeding it with whelpes cattes ratts & the like. in all that while not giving it any water wch I afterwards presented unto the [colledge of physitians at London struck out] my worthy friend Dr Scarburgh.

[3] The "Aquila" of Gesner here referred to is evidently the Golden Eagle, which species Browne is careful to mention that he had not met with in this

county, and that the specimen he sent to Dr. Scarburgh, more than once mentioned, was brought from Ireland. This bird has never been recorded alive in Norfolk. Immature White-tailed Eagles, the "Halyaetus" of the text, still occur almost every autumn or winter on this coast, but no mature example has hitherto been killed. Browne's friend, Sir Charles Scarburgh (1616-1694), was born in London, and is buried at Cranford, in Middlesex. He seems to have been greatly distinguished as an anatomist and physician. He was a friend of William Harvey, whom he succeeded as Lumleyan Lecturer at the College of Physicians (of which he was elected a fellow in 1650). Harvey, out of regard for his "lovinge friend" Dr. Scarburgh, bequeathed to him his "little silver instruments of surgerie" and his velvet gown. ("Dict. of Nat. Biog.") The Golden Eagle sent him by Browne was kept in the College of Physicians in Warwick Lane for two years.

of other sorts of Agles there are severall kinds especially of the Halyaetus or fenne Agles some of 3 yards & a quarter from the extremitie of the wings. whereof one being taken aliue grewe so tame that it went about the yard feeding on fish redherrings flesh & any offells without the least trouble.

There is also a lesser sort of Agle called an ospray[4] wch houers about the fennes & broads & will dippe his [foot crossed out] claws & take up a fish oftimes for wch his foote is made of an extraordinarie roughnesse for the better fastening & holding of it & the like they will do unto cootes.

[4] This species is a not unfrequent autumn visitor to the Broads and Rivers of Norfolk. Browne names it correctly, but there was much confusion with regard to this species in the minds of the old authors. Willughby knew the bird and calls it the "Bald Buzzard," but in describing its nesting site and eggs (probably not on his own authority,) evidently confounds it with the Marsh Harrier, for he says that "it builds upon the ground among reeds, and lays three or four large white eggs of a figure exactly elliptical, lesser than hens' eggs." See Note 6.

[Fol. 6.] Aldrovandus takes particular notice of the great number of Kites[5] about London & about the Thames. wee are not without them heare though not in such numbers. there are also the gray & bald Buzzard[6] [wch the all wth crossed out] of all wch the great number of broad waters & warrens makes no small number & more than in woodland counties.

[5] The Glede, or Puttock, of Turner, once so plentiful, is now only an extremely rare visitor to Norfolk. In 1815, it appears from Hunt ("British Ornithology"), not to have been uncommon, but the same authority in his list of Norfolk Birds contributed to Stacey's "History" of that County, speaks of the Kite as having in 1829 become extremely rare. It probably ceased to nest in this County about the year 1830, or perhaps a little later. Browne's reason for its comparative scarcity about the City of Norwich, viz., the abundance of Ravens mentioned at p. 27 infra, is very interesting to us in the present day when Kites and Ravens are almost equally rare.

[6] It seems likely that Browne here refers to two species of Harrier, the Grey Buzzard being the male of the Hen Harrier (including of course Montagu's Harrier which was not discriminated till long after) in its grey adult plumage, whereas the Marsh Harrier, with its light yellow head, to which the word "bald" as then used might well be applied, would stand for the "Bald Buzzard." The Harriers, which were till long after the time he wrote extremely numerous, are generally called "Buzzards" by the natives, and it will be noticed at p. 15 infra, that what is doubtless intended for the Marsh Harrier is spoken of as an enemy to the Coots; also at p. 56, it is said that young Otters "have been found in the Buzzards nests," a very likely circumstance with so fierce a bird, and one of which I have an impression I have heard in recent years. The Hen Harrier is now an extremely rare bird with us; the Marsh Harrier still occasionally nests in the Broads, and Montagu's Harrier now and then attempts to rear a brood, but even should the parents succeed in escaping it is very seldom they carry their young with them. Professor Newton has kindly favoured me with the following additional interesting note on this bird. "The Marsh Harrier is certainly the 'Balbushardus' of Turner (1544), which, though he says it is bigger and longer than the ordinary Buteo, has a white patch on the head and is generally of a dark brown (fuscus) colour, hunting the banks of

rivers, pools, and marshes, living by the capture of Ducks, and the black birds which the English call Coots (Coutas). This he, Turner, has himself very often seen, and he describes its habits correctly; adding that it also takes Rabbits occasionally. Gesner, 1555, quotes Turner, but refers the Bald Buzzard to the Osprey (which he figures), and so the mistake began. Certainly Willughby's Bald Buzzard is the Osprey, but his book was not published when Browne wrote."

Cranes[7] are often seen here in hard winters especially about the champian & feildie part it seems they have been more plentifull for in a bill of fare when the maior entertaind the duke of norfolk I meet with Cranes in a dish.

[7] In the present day the Crane is only a rare straggler to this country generally at the seasons of its migration; that it was in times past abundant in suitable localities there is ample evidence; that it also bred in the fens of the Eastern Counties there is no reason to doubt, but very little direct evidence is forthcoming, therefore every fact bearing upon this point is of value. Had Sir Thomas Browne written with the intention of publishing his observations he would doubtless have told us much about this grand bird, which would have been of the greatest interest to modern ornithologists, but even the above brief remarks, as will be seen, are worthy of note.

With regard to the occurrence of the Crane in the fens of East Anglia we have the following evidence; its fossil remains have been found in the peat at Burwell, in Cambridgeshire, and in excavating the docks at Lynn. Turner, in his "Avium Historia," Coloniae, 1544, speaks of having seen young Cranes in this country, and as he passed fifteen years at Cambridge, it was probably in that neighbourhood that he met with them; then again there is the Act of Parliament, passed in 1534 (25th Hen. VIII. c. ii.), prohibiting the taking of their eggs (amongst those of other species) under a penalty of twenty pence. All this is well known, but being desirous to ascertain whether any reference to the Crane was to be found in the records of the Corporation of Norwich, Mr. J. C. Tingey, F.S.A., the custodian of the Muniment Room, at my request, most kindly searched the accounts of the City Chamberlain between the years 1531 and 1549. He there found numerous entries of sums expended in the purchase of cranes, swans, porpoises, &c., as presents to the Dukes of Norfolk and Suffolk and others, and amongst them, on the 6th of June, 1543, a charge for a "yong pyper crane" from Hickling, which appears conclusive evidence of the breeding of this bird near Norwich at that time. (See "Transactions of the N. and N. Nat. Soc.," vii., pp. 160-170.)

In Wilkin's Edition of the Notes the statement, "I met" with Cranes in a dish should be, "I meet with," &c., as it is in the original. The occasion referred to was probably an entertainment given by the Mayor of Norwich, on the Guild day in 1663, which in that year fell on the 19th June; at this banquet Henry, Duke of Norfolk and the Hon. Henry Howard were present, and the latter presented to the City a silver basin and ewer of the value of L60. Can it be that even at that time young Cranes were to be obtained? otherwise the middle of June seems a most unseasonable time for such a dish; for in a copy of a curious old manuscript, dated 1605, and published in the 13th Volume of "Archaeologia" (p. 315), entitled "A Breviate touching the Order and Government of a Nobleman's house," &c., there is a "Monthlie Table, for a Diatorie" for each month in the year, and the Crane appears only in the tables from November till March inclusive. The modern gourmet would view with disgust some of the dishes included in this "diatorie" if set before him--only to mention among birds, auks, stares, petterells, puffines, didapers, and martins. The crane being "in the dish" must not be subjected to the vulgar process of "kervyng," but in the stilted heraldic language of the day must be "desplayed," whereas a heron must be "dismembered" and a bittern "unjointed." The price of a crane varied from 3s. 4d. to 5s., and a fat swan from 3s. to 4s. The sum of 6d. mentioned in the le Strange Household-book, in the year 1533 (see "Archaeologia," vol. xxv., p. 529), quoted in Yarrell's "British Birds," iii., p. 180, was only the reward for bringing in a crane killed on the estate. That Cranes must at times have been numerous in Norfolk in the sixteenth century is evident, for in an account of the presents sent to William Moore, Esq., of Loseley, on the occasion of the marriage of his daughter, on 3rd November, 1567, Mr. Balam, "out of Marshland in Norfolk," sent him nine cranes, nine swans, and sixteen bitterns, with a large number of other wild-fowl.

In hard winters elkes[8] a kind of wild swan are seen in no small numbers. in whom & not in co[=m]on swans is remarkable that strange recurvation of the windpipe through the sternon. & the same is also obseruable in cranes. tis probable they come very farre for all the northern discouerers have [ha struck out] obserued them in the remotest parts & like diuers [&] other northern birds if the winter bee mild they co[=m]only come no further southward then scotland if very hard they go lower & seeke more southern places. wch is the cause that sometimes wee see them not before christmas or the hardest time of winter.

[8] The "Elke" is an obsolete name for the Wild Swan (Cygnus musicus), which occurs in the present day in the same numbers and under precisely similar circumstances as Browne describes; but of course this was the only species of wild swan known to him. The remarkable recurvation of the trachea within the keel of the sternum, which also prevails to a greater or less degree in four out of the five or six species of Cygnus found in the Northern Hemisphere, did not escape Browne's notice, although he was not the first to describe it, and he rightly observes that this peculiarity is absent in the Mute Swan (C. olor), but exists in a different and even more exaggerated form in the Crane. He, however, was mistaken as to the extreme northerly range which he assigns to this species. So marked a feature as the absence of the "berry" on the beak of this species did not escape Browne's observation, and he refers to it in the eighth letter to Merrett, who in his second letter to Browne remarks "the difference in the elk's bill by you signified is remarkable to distinguish it from others of its kind," indicating that this distinction was previously unknown to him.

A white large & strong billd fowle called a Ganet[9] which seemes to bee the greater sort of Larus. whereof I met with one kild by a greyhound neere swaffam another in marshland while it fought & would not bee forced to take wing another intangled in an herring net wch taken aliue was fed with herrings for a while it may be named Larus maior Leucophaeopterus as being white &

the top of the wings browne.

[9] As a rule the Gannet does not approach the shore, except to breed, but follows the shoals of fish far out at sea. The circumstance mentioned by Browne is by no means singular, and several such instances of storm-driven Gannets being captured far inland are recorded. The "Scotch Goose, Anser scoticus," mentioned further on (p. 13 infra), is also in all probability intended for the Gannet; it is the Anser Bassanus sive Scoticus of Jonston. The "Marshland" here mentioned is a tract of country reclaimed in ancient times from the sea, lying to the west of the town of Lynn, of some 57,000 acres in extent, and bordering upon the estuary of the Wash.

[Fol. 7.] In hard winters I have also met with that large & strong billd fowle wch clusius describeth by the name of Skua Hoyeri[10] [fr struck out] sent him from the faro Island by Hoierus a physitian. one whereof was shot at Hickling while 2 thereof were feeding upon a dead horse.

[10] Willughby ("Ornithology," English Ed., p. 348) gives a good description of the Great Skua (Stercorarius catarrhactes) under the name of Catarracta, a skin of which he says was sent him by Dr. Walter Needham, and rightly identified it with the Skua which Hoier sent to Clusius, but his figure is evidently drawn from a skin of the Great Black-backed Gull. Hoier, whose name so often occurs about this time in connection with birds from the north, was a physician, living at Bergen in Norway. The Great Skua still breeds in sadly reduced numbers on the Shetland and Faroee Islands, but is rarely met with in Norfolk.

As also that [strong struck out] large & strong billd fowle [Clusius nameth struck out] spotted like a starling wch clusius nameth Mergus maior farroensis[11] as frequenting the faro islands seated above shetland. one whereof I sent unto my worthy friend Dr Scarburgh.

[11] The bird here mentioned is doubtless the Great Northern Diver, Colymbus glacialis. In another place Browne again refers to it as Mergus

maximus Farrensis, which Clusius ("Exotic.," p. 102) calls Mergus maximus Farrensis, a name used by Willughby as a synonym for his "Greatest Speckled Diver or Loon" (p. 341). This bird is known to our fishermen as the Herring Loon, the Red-throated and perhaps also the Black-throated Divers being called Sprat Loons. It is a pity Browne's "draught" is not forthcoming.

Here is also the pica marina[12] or seapye many sorts of Lari,[13] seamewes & cobs. the Larus maior in great abundance [about struck out] in [written above] herring time about yarmouth.

[12] The Oyster Catcher, or Sea Pie, is found in greater numbers on the north-west portion of the County of Norfolk than on the eastern shore; it breeds occasionally about Wells, where it is universally known as the "Dickey-bird."

[13] Browne here refers to the family in general terms. The various species of Gulls in their different stages of plumage were very puzzling to the Ornithologists of the last century, and it is often extremely difficult to say to what individual species they refer. By Larus major he would probably mean the Black-backed and Herring Gulls which are found on the shore all the year round, most frequently in the immature plumage, but they most abound "in herring time." By far the commonest species at all times is Browne's Larus alba or Puet, the Black-headed Gull. Large flocks of this species and L. canus frequent Breydon and the tidal shores, especially the young birds of the year. There are now two large breeding-places of the Black-headed Gull in Norfolk, a very old-established one at Scoulton Mere, and a more recent colony at Hoveton Broad. The former extensive gullery at Horsey, mentioned by Browne, has long since been banished by the drainage of the marsh they frequented, and it is probable that a small colony which bred on Ormesby Broad some forty years ago, owed its origin to their banishment from Horsey. They, in their turn, deserted Ormesby on the erection of the works for supplying Yarmouth with water about the year 1855, and fixed upon Hoveton as their new home, in which place, as at Scoulton, they are carefully preserved.

Professor Newton has been kind enough to furnish me with the following note on the Terns. "Larus cinereus of Aldrovandus (and afterwards of Jonston), is said to be of three kinds: one with red legs, apparently the Black-headed Gull, and figured by Jonston, the second with yellow legs and a slender curved black bill, the third with a pointed scarlet bill. Both these last were most likely Terns--and all these were grey above and white below. Gesner quotes Turner for Sterna, and there is no doubt that his bird of that name was a Black Tern; but Gesner says that it is the Stirn of the Frisians, and figures a white and grey bird with a black head only (most likely a Common Tern, but possibly one of the larger species), as Sterna, thus using the word in a more general sense, and it may have been so used in Browne's time. I see no impossibility in people having thought of eating Terns in those days [as to that see Note 7, p. 6 ante]. The Common Tern was most likely very abundant, and we know that the Black Tern was exceedingly common in certain reed-beds, as stated by Turner, and noisy beyond measure." The Great and Lesser Terns still nest in one or two localities on our coast, although as the result of great persecution in very reduced numbers. The Black Tern, or Mire Crow, has quite ceased to do so.

Larus alba or puets in such plentie about Horsey that they sometimes bring them in carts to norwich & sell them at small rates. & the country people make use of their egges in puddings & otherwise. great plentie thereof haue bred about scoulton [mere struck out] meere, & from thence sent to London.

Larus cinereus greater & smaller, butt a coars meat. commonly called sternes.

Hirundo marina or sea swallowe a neat white & forked tayle bird butt longer then a swallowe.

The ciconia or stork[14] I have seen in the fennes & some haue been shot in the marshes between this and yarmouth. [See also third letter to Merrett and Appendix D.]

[14] Although it has been met with in Norfolk, more frequently than perhaps in any other part of England, the Stork was never other than a rare spring and

autumn visitor to Norfolk. Turner writes of it in 1544 as unknown in England, save as a captive, and Merrett a hundred years later says it rarely flies hither, which is equally true at the present time. Hewittson ("Eggs of Brit. Birds," Ed. 3, ii., p. 309; under Crane) was evidently misled by some remarks made by Evelyn, who visited Sir Thomas Browne in Norwich in October, 1671, and says in his diary that he saw Browne's "Collection of the eggs of all the fowl and birds he could procure; that country, especially the promontory of Norfolk, being frequented, as he said, by several birds which seldom or never go further into the land--as cranes, storks, eagles, and a variety of water-fowl." From this Hewitson infers that the Stork bred in Norfolk, a construction which the somewhat ambiguously worded passage will certainly not bear. I imagine collections of eggs were not very common in Browne's time.

[Fol. 8.] The platea or shouelard,[15] wch build upon the topps of high trees. they haue formerly built in the Hernerie at claxton & Reedham now at Trimley in Suffolk. they come in march & are shot by fowlers not for their meat butt the handsomenesse of the same, remarkable in their white colour copped crowne & spoone or spatule like bill.

[15] This interesting record has recently been supplemented by a much earlier record of the breeding of the "Popeler," or Shovelard, in Norfolk. Professor Newton ("Transactions of N. and N. Nat. Soc.," vi., p. 158) has called attention to an ancient document bearing date A.D. 1300, instituting a commission to inquire into the harrying of the eyries of these and other birds, &c., at Cantley and other places in Norfolk. Documents also exist, showing that in 1523 they nested at Fulham in Middlesex, and in 1570 in West Sussex, as pointed out by Mr. Harting in the "Zoologist" for 1877, p. 425, and 1886, p. 81, in each case constructing their nests in trees. At what precise date this bird ceased to breed in Norfolk and Suffolk is unknown, but Sir T. Browne's statement that they were "shot by fowlers not for their meat, butt the handsomenesse of the same," probably explains the circumstances which brought about that event. The Spoonbill visits Norfolk regularly every spring in small parties now more numerously than a few years since, which possibly may be accounted for by the destruction of nearly all its breeding-places in

Holland, and it is possible that with due encouragement it might again be induced to breed in some of the localities in the Broads still suitable for the purpose.

corvus marinus. cormorants.[16] building at Reedham upon trees from whence King charles the first was wont to bee supplyed. beside the Rock cormorant wch breedeth in the rocks in northerne countries & cometh to us in the winter, somewhat differing from the other in largenesse & whitenesse under the wings.

[16] The Cormorant continued to nest in the trees on the shore of Fritton Lake for many years after Sir T. Browne's time. A manuscript note in a copy of Berkenhout's "Natural History of Great Britain and Ireland," published in 1769, is descriptive of a Cormorant killed at Belton Decoy (near the same lake) on the 11th September, 1775, and also states that "a vast number of these birds, even to some thousands, roost every night upon the trees," being in the neighbourhood of the decoy they are never shot, and "build their nests upon the top of these trees." According to Mr. Lubbock ("Fauna of Norf.," Ed. 2, p. 174), "in 1825 there were many nests at Herringfleet, also on Fritton Lake, and in 1827 not one." We may therefore assume that they ceased to nest at Herringfleet in 1825 or 1826. It will be noticed that Browne made free use of young Cormorants in his experiments as to the properties of certain drugs (cf. Wilkin, iv., p. 452), which would seem to indicate that he could obtain a plentiful supply of these birds. When the Cormorants ceased to breed at Reedham is unknown. They are not unfrequently seen now, generally in spring and autumn. The Rock Cormorant was possibly the Crested Cormorant or Shag.

A sea fowl called a shearwater,[17] somewhat billed like a cormorant butt much lesser a strong & feirce fowle houering about shipps when they [clense struck out] cleanse their fish. 2 were kept 6 weekes cra[=m]ing them with fish wch they would not feed on of themselues. the seamen told mee they had kept them 3 weekes without meat. & I giuing ouer to feed them found they liued 16 dayes without [any hin struck out] taking any thing.

[17] Willughby's first acquaintance with the adult Manx Shearwater ("Ornithology," p. 334) was from a drawing sent him by Sir T. Browne, who describes the bird, as above, under the accepted name of Shearwater, and Willughby's excellent figure on plate lxvii. (which plate I believe is not to be found in some copies of the "Ornithology," and to which there is no reference in the text) has all the appearance of having been drawn from life. The drawing here referred to is mentioned by Ray in his "Collection of English words not generally known," as having been received, with others, from the "learned and deservedly famous Sir Thomas Browne, of Norwich." George Edwards ("Gleanings of Nat. Hist.," vii., p. 315), prior to 1764. says that he went to the British Museum and examined Browne's "old draught," but I could not find it among any of the papers I examined. In Browne's fourth letter to Merrett, by an error in the transcription, he is made by Wilkin to say that he kept twenty of these birds alive for five weeks; in the MS. it is clearly only two.

Barnacles[18] Brants Branta [wer struck out] are co[=m]on

[18] Barnacle and Brent Geese as we know them, the first by no means common here; the Wild Goose, probably Anser cinereus; the Scotch Goose (see Note 9), probably the Gannet; and the Bergander, an old name for the Sheld-drake, as used by Turner in 1544, and derived from the Dutch Berg-eende, German Bergente ("Dict. Birds," p. 835). Browne's statement that this bird formerly bred about Northwold, or as it is even now occasionally called by the natives, "Norrold," some twenty miles from the sea; or, as he says, in the fourth letter to Merrett, "abounding in vast and spatious commons," is very interesting, although not a solitary instance, for I am informed that this bird breeds in the present day on the Gull Lake, Twig Moor, in Lincolnshire; but that it should have chosen such a nesting site is not more surprising than the fact of the Ring Plover, quite as strictly a marine species, frequenting the extensive sandy warrens about Thetford and Brandon, near the south-west border of the county, for the same purpose, as they still continue to do. But for Browne's mention of the circumstance we should not have been aware of this

singular departure from the normal nesting habits of the Sheld-duck, as no tradition I believe exists on the subject, and at present it only nests in the sand-hills in some parts of the coast of N.W. Norfolk.

sheldrakes sheledracus jonstoni

Barganders a noble coloured fowle vulpanser wch breed in cunny burrowes about norrold & other places.

[Fol. 9.] Wild geese Anser ferus.

scoch goose Anser scoticus.

Goshander,[19] merganser.

[19] This evidently refers to the Goosander, which as he says in another place most answers to the Merganser.

Mergus acutirostris speciosus or Loone an handsome & specious fowle cristated & with diuided finne feet placed very backward and after the manner of all such wch the Duch call [Assf struck out] Arsvoote.[20] they haue a peculiar formation in the leggebone wch hath a long & sharpe processe extending aboue the thigh bone [it struck out] they come about April & breed in the broad waters so making their nest on the water that their egges are seldom drye while they are sett on.

[20] This well describes the Great-crested Grebe, which Browne rightly says comes to us about the month of April. Browne notices the peculiar formation of the tibia in this family of birds, but it had long been known. The next, named Mergus acutirostris cinereus, is most likely the same species in winter plumage. The other birds mentioned are Mergus minor, the Little Grebe or Dabchick, and M. serratus, the Red-breasted Merganser, even now known as the "Saw-bill."

Mergus acutarostris cinereus [another d struck out] wch seemeth to bee a difference of the former.

Mergus minor the smaller diuers or dabchicks in riuers & broade waters.

Mergus serratus the saw billd diuer bigger & longer than a duck distinguished from other diuers by a notable sawe bill to retaine its slipperie pray as liuing much upon eeles whereof we haue seldome fayled to find some in their bellies.

Diuers other sorts of diuefowle more remarkable the mustela fusca & mustela variegata[21] the graye dunne & the variegated or partie coloured wesell so called from the resemblance it beareth vnto a wesell in the head.

[21] The Smew, male and female, or either in the immature plumage are here referred to.

[Fol. 12.[I]] many sorts of wild ducks[22] wch passe under names well knowne unto the fowlers though of no great signification as smee [wige struck out] widgeon Arts ankers noblets.

[I] Fols. 10 and 11 are (10 written on both sides) on the "Ostridge," vide Wilkin, Vol. 4, p. 337-9. The paper is a different size, 11-1/2 by 7-1/2, and the article is evidently bound out of place.

[22] The local names of the various Ducks are simply legion and differ both in time and place, not to mention the confusion occasioned by sex and season when these birds were not so well understood as at present. Many such names are quite lost, as "Ankers" and "Noblets," but the following are a few examples: Adult Smew, White Nun; female or immature Smew, Wesel Coot; the Wigeon was known as the Smee, Whewer, or Whim; the Tufted Duck, Arts or Arps; the Gadwall, Grey Duck or Rodge; the Pochard, Dunbird; the Shoveller, Beck or Kertlutock (Hunt); Pintail, Sea Pheasant or Cracker; Long-tailed Duck, Mealy Bird; Golden Eye, Morillon or Rattle-wing; Scaup, Grey-back, and on Breydon White-nosed Day Fowl; Scoter, Whilk; Velvet Scoter, Double Scoter

(Hunt); Teal, Crick; Garganey, Summer Teal, Pied Wigeon, Cricket Teal; other names might be mentioned, and some will be found in the notes which will follow. Anas platyrhincus here mentioned is the Shoveller. It may seem strange that the abundance of Teal should in any way be attributed to the number of Decoys, but such was really the case, the quiet and shelter afforded by these extensive preserves being very favourable to the increase of all the members of the Duck family, especially to those breeding in their immediate neighbourhood. In the returns of the old Decoys, Teal figured largely; in the present day they form a very much smaller proportion of the spoils.

the most remarkable are Anas platyrinchos [sic] a remarkably broad bild duck.

And the sea phaysant holding some resemblance unto that bird [in the tayle crossed out] in some fethers in the tayle.

Teale Querquedula. wherein scarce any place more abounding. the condition of the country & the very many decoys [mo struck out] especially between Norwich and the sea making this place very much to abound in wild fowle.

fulicae cottae cootes[23] in very great flocks upon the broad waters. upon the appearance of a Kite or buzzard I have seen them vnite from all parts of the shoare in strange numbers when if the Kite stoopes neare them they will fling up [and] spred such a flash of water up with there wings that they will endanger the Kite. & so [es struck out] keepe him of [in of struck out] agayne & agayne in open opposition. & an handsome prouision they make about their nest agaynst the same bird of praye by bending & twining the rushes & reeds so about them that they cannot stoope at their yong ones or the damme while she setteth.

[23] In the present day the Coots have nothing to fear from Kites and little from Moor Buzzards; it may be that it is in consequence of this that they have discontinued the practice of twining the rushes and reeds above their nests in the manner mentioned above as being an unnecessary precaution. I have,

however, in some cases noticed some approach to this practice. The Coot, although fairly numerous on the Broads, appears to be far less so than formerly. Lubbock, in his "Fauna of Norfolk," says on asking a Broadman how many Coots there were on Hickling Broad, his reply was, "About an acre and a half," referring to their practice of swimming evenly at regular distances from each other without huddling together in dense masses, like wild-fowl.

I am indebted to Professor Newton for the following additional note on the Coot. He says "Turner, and after him Gesner, was puzzled as to what was the Fulica of classical writers (Virgil and others), and thought it to be some kind of Gull; but the Fulica of later authors was certainly the Coot, as shown by Gesner's figure."

Gallinula aquatica[24] more hens.

[24] Moor-hens are of course numerous in all suitable localities, and the Water Rail is still fairly common, but its eggs have a market value and are (or were) sadly stolen; a few years ago a London dealer is said to have received over 200 eggs of this bird in one season from Yarmouth.

And a kind of Ralla aquatica or water Rayle.

[Fol. 13.] An onocrotalus or pelican[25] shott upon Horsey fenne 1663 May 22 wch stuffed and cleansed I yet retaine it was 3 yards & half between the extremities of the wings the chowle & beake answering the vsuall discription the extremities of the wings for a spanne deepe browne the rest of the body white. a fowle [not found struck out] wch none could remember upon this coast. about the same time I heard one of the kings pellicans was lost at St James', perhaps this might bee the same.

[25] There is every reason to believe that a species of Pelican, probably from its size P. crispus, was formerly an inhabitant of the East Anglian Fens; its bones have been found in the peat on three occasions, one of these being the bone of a bird so young as to show that it must have been bred in the locality,

and therefore that the species was a true native and not a casual visitant. Bones of a species of Pelican have also been found in the remains of lake-dwellings at Glastonbury, in Somersetshire.

With regard to the species of the bird recorded by Browne and its origin, he is careful to point out that a Pelican had about that time escaped from the King's collection in St. James' Park, and to surmise that it might be the same bird; from what follows this seems probable, but as P. onocrotalus is believed to stray occasionally into the northern parts of Germany and France ("Dict. of Birds," p. 702) the occurrence of that species on the East Coast of Britain, where, even at present, it would find a state of things in every way suited to its requirements (guns excepted), would not be very extraordinary. Browne's Pelican was killed in May, 1663, and although Dr. Edward Browne visited St. James' Park in February, 1664, and saw "many strange creatures," including the Stork with the wooden leg (mentioned by Evelyn), he says nothing of the Pelicans, still it may be that it was from him that his father heard of the escape. Evelyn, in his Diary, mentioned that he visited St. James' Park on February 9th, 1665, and speaks of only one Pelican, which he states was brought from Astrakan by the Russian Ambassador as a present to the King; Willughby says distinctly that the Emperor of Russia sent the King two Pelicans, and further, that he took the description in his "Ornithology" from a bird in the Royal Aviary, St. James' Park, near Westminster; it seems therefore highly probable that Browne's bird was one of these which had escaped from confinement. But a rather curious circumstance arises out of this, the bird described by Willughby does not appear to be P. onocrotalus, but a similar species, P. roseus, found chiefly in Indio-China and westward to South-eastern Europe, but occurring as far west as the River Volga ("Cat. of Birds," B. M., xxvi., p. 466). In this Mr. Ogilvie Grant, the author of that section of the Catalogue, whom I consulted, agrees with me, and the locality whence the birds were derived, mentioned by Willughby, renders not unlikely. Onocrotalus in Browne's time was a general term for "the Pelican," and he probably knew but one species and one individual, the escaped bird from Charles II.'s Aviary. Browne's very miscellaneous collection was destroyed by the authorities at the time of the plague (see ninth letter to Merrett), and probably the remains of this Pelican perished with the rest.

Anas Arctica clusii wch though hee placeth about the faro Islands is the same wee call a puffin co[=m]on about Anglisea in wales & sometimes [for struck out] taken upon our seas not sufficiently described by the name of puffinus the bill being so remarkably differing from other ducks & not horizontally butt meridionally formed to feed in the clefts of the rocks of insecks, shell-fish & others.

The great number of riuers riuulets & plashes of water makes hernes [to abound in these struck out] & herneries to abound in these parts. yong hensies being esteemed a festiuall dish & much desired by some palates.

The Ardea stellaris botaurus, or bitour[26] is also co[=m]on & esteemed the better dish. in the belly of one I found a frog in an hard frost at christmas. another I kept in a garden 2 yeares feeding it with fish mice & frogges. in defect whereof making a scrape for sparrowes & small birds, the bitour made shifft to maintaine herself upon them.

[26] This is one of the birds once common enough in Norfolk, which in the present day is only a winter and spring migrant. The last eggs of the Bittern were taken in this county on 30th of March, 1868; the last "boom" of a resident was heard in May, 1886, in the August of which year a young female was killed at Reedham with down still adhering to its feathers; this was probably the last Norfolk-bred Bittern. In the "Vulgar Errors," book 3, chapter xxvii., section 4, is a discourse on the "mugient noise" of the Bittern and the mode of its production, and in a foot-note in the same place is a curious anecdote illustrating the difficulty of detecting a wounded Bittern, even when marked down in short, recently mown grass and flags. The spring cry of the Bittern is mentioned by Robert Marsham in his unpublished journal nineteen times, between the years 1739 and 1775, as first heard at Stratton Strawless, generally between the 15th of March and the 15th of April; and it was on the 14th of the latter month that Benjamin Stillingfleet records it in the "Calendar of Flora" as heard in the same locality in 1755. He does not describe the note,

but uses the words "makes a noise." Marsham, however, on one occasion, in 1750, a very early year, records it on the 20th of February. As a once familiar sound, but one which will probably never again be heard here under purely normal conditions, these dates seem worthy of recording.

[Fol. 14.] Bistardae or Bustards[27] are not vnfrequent in the champain & feildie part of this country a large Bird accounted a dayntie dish, obseruable in the strength of the brest bone & short heele layes an egge much larger then a Turkey.

[27] The last of the Norfolk and therefore certainly the last of the British-bred Bustards, was killed in May, 1838; those which have since occurred in this country were Continental immigrants. An exhaustive history of the extinction of this bird will be found in Stevenson's "Birds of Norfolk," vols. 2 and 3. The Bustard, although found in some numbers, associated in small flocks or "droves" in the few localities which it frequented in Great Britain, was probably never a very numerous species. The following extract from one of Browne's letters to his son Edward, dated April 30th, and written probably in 1681, shows that he was on the verge of discovering an anatomical peculiarity in this family of birds, which in after years gave rise to much controversy. He says, "yesterday I had a cock Bustard sent me from beyond Thetford. I never did see such a vast thick neck: the crop was pulled out, butt as [a] turkey hath an odde large substance without, so hath this within the inside of the skinne, and the strongest and largest neckbone of any bird in England. This I tell you, that if you meet with one you may further observe it." The presence of a gular pouch in the Bustard was first demonstrated by James Douglas, a Scotch Physician, in 1740, and it appears to be fully developed only in the adult male bird, and at the breeding season. Hence, although it has undoubtedly been found on several occasions, the frequent unsuccessful searches for it under unfavourable conditions led to much scepticism as to its existence. The use of this singular appendage is still a moot point, but it seems probable that it has to do with "voice production," and assists in the remarkable "showing off" exhibited by the male bird in the breeding season. Pennant, in his "British Zoology," 1768, i., p. 215, gives a sentimental account of its use, and an

exaggerated estimate of its proportions. In the Tables of Dietary referred to at p. 6 (note) ante, the Bustard is mentioned as in season from October to May.

Morinellus or Dotterell[28] about Thetford & the champain wch comes vnto us in september & march staying not long. & is an excellent dish.

[28] The Dotterel visits us much as in Sir T. Browne's time, but in decreased numbers. The Sea Dotterel which Wilkin supposes to be the Ring Plover, is undoubtedly the Turnstone. Willughby says, "Our honoured Friend, Sir Thomas Browne, of Norwich, sent us the picture of this bird by the title of the Sea Dotterel." This is also mentioned in the fifth letter to Merrett. See "Birds of Norfolk," ii., p. 82, for an interesting account of Dotterel hawking near Thetford by James I. in the year 1610.

There is also a sea dotterell somewhat lesse butt better coloured then the former.

Godwyts taken chiefly in marshland, though other parts not without them accounted the dayntiest dish in England & I think for the bignesse, of the biggest price.

Gnatts or Knots [only so far on p. 14, but as follows on fol. 13 verso].

Gnats or Knots a small bird which taken with netts grow excessively fatt. If [by mew struck out] being mewed & fed with corne a candle lighted in the roome they feed day & night, & when they are at their hight of fattnesse they beginne to grow lame & are then killed or [else they will fall aw struck out] as at their prime & apt to decline.

[resume p. 14.] Erythropus or Redshanck a bird co[=m]on in the marshes & of co[=m]on food butt no dayntie dish.

A may chitt[29] a small dark gray bird litle bigger then a stint of fatnesse beyond any. it comes in may into marshland & other parts & abides not aboue a moneth or 6 weekes.

[29] Mr. Stevenson, "Birds of Norfolk," ii., p. 233, gives his reasons for coming to the conclusion that the Sanderling (Calidris arenaria) is here referred to, which the absence of a hind toe (see third letter to Merrett) tends to confirm. The "Churre" is only a variant of the name "Purre," by which the next species, the Stint, is commonly known, and the Green Plover, now applied to the Lapwing, is an old name for the Golden Plover, which he rightly says [p. 20] does not breed in Norfolk.

[fol. 13 verso.] Another small bird somewhat larger than a stint called a churre & is co[=m]only taken amongst them.

[resume fol. 14.] Stints in great numbers about the seashore & marshes about stifkey Burnham & other parts.

Pluuialis or plouer green & graye in great plentie about Thetford & many other heaths. they breed not with us butt in some parts of scotland, and plentifully in Island [Iceland].

[Fol. 15.] The lapwing or vannellus co[=m]on ouer all the heaths.

Cuccowes[30] of 2 sorts the one farre exceeding the other in bignesse. some have attempted to keepe them in warme roomes all the winter butt it hath not succeeded. in their migration they range very farre northward for in the summer they are to bee found as high as Island.

[30] The circumstance which gave rise to the idea that there were two kinds of Cuckoos, differing only in size, might possibly be discovered were it worth the research; possibly it would be found that the second species was of foreign origin. Aldrovandus, as quoted by Willughby, says, "Our Bolognese Fowlers do unanimously affirm, that there are found a greater and a lesser sort of Cuckows; and besides, that the greater are of two kinds, which are distinguished one from the other by the only difference of colour: but the

lesser differ from the greater in nothing else but magnitude." Perhaps it was Browne's latent respect for antiquity which led him to mention the tradition.

Avis pugnax. Ruffes[31] a marsh bird of the greatest varietie of colours euery one therein somewhat varying from other. The female is called a Reeve without any ruffe about the neck, lesser then the other & hardly to bee got. They are almost all cocks & putt together fight & destroy each other. & prepare themselues to fight like cocks though they seeme to haue no other offensive part butt the bill. they loose theire Ruffes about the Autumne or beginning of winter as wee haue obserued [they struck out] keeping them in a garden from may till the next spring. they most abound in Marshland butt are also in good number in the marshes between norwich & yarmouth.

[31] It is only necessary to add to Browne's interesting account of this remarkable bird that it lingered longer in Norfolk as a breeding species than in any other part of Britain, but that although it still visits us in spring it is doubtful whether it has bred for the last few years in the one favourite locality to which it clung so tenaciously. The "Marshland," here referred to as explained in a previous note, is a tract of country situated in north-west Norfolk, near King's Lynn.

Of picus martius[32] or woodspeck many kinds. The green the Red the Leucomelanus or neatly marked [red crossed out] black & white & the cinereus or dunne calld [a re struck out] little [bird calld written above] a nuthack. remarkable in the larger are the hardnesse of the bill & skull & the long nerues wch tend vnto the tongue whereby it strecheth out the tongue aboue an inch out of the mouth & so [lik crossed out] licks up insecks. they make the holes in trees without any consideration of the winds or quarters of heauen butt as the rottenesse thereof best affordeth conuenience.

[32] Picus martius is here used, as it is by Sibbald, and all preceding writers, in a general sense for all birds commonly called "Woodpeckers," and does not imply that the Great Black Woodpecker (Picus niger maximus, of Ray's Synopsis), to which species the name was restricted by Linnaeus, is found here,

and Browne goes on to mention the three British Woodpeckers, the Green, the Red, by which the Great Spotted Woodpecker is intended, and the Leucomelanus, or Lesser-spotted Woodpecker. He also includes the Nuthatch, which was at that time (as well as the Wryneck) called a "Woodpecker." In this passage Browne, in making a correction, does not seem to have proceeded far enough, the word which Wilkin has rendered "dun-coloured," is certainly "dunne calld" in the MS.; but there are two alterations in the passage, and there is little doubt that he intended to write "dunne cull'd" (or coloured), which would make it read as Wilkin has printed it. The use of the word "nerve," for tendon or ligament, was in accordance with the phraseology of the time.

[fol. 15 verso.] black heron[33] black on both sides the bottom of the neck neck [sic] white gray on the outside spotted all along with black on the inside a black coppe of small feathers some a spanne long. bill poynted and yallowe 3 inches long

[33] This passage is not part of the original MS., but is written on a separate slip of paper and pasted on the left-hand side of the opening (p. 15 verso). I doubt whether it is more than a casual memorandum, descriptive possibly of the plumage of the Purple Heron, but not intended to apply to any Norfolk bird. The Black Heron of Willughby is the Glossy Ibis, a bird which is said to have been known to the West Norfolk gunners as the "Black Curlew."

back heron coloured intermixed with long white fethers

the flying (?) fethers black

the brest black & white most black

the legges & feet not green but an ordinarie dark cork [?] colour.

[Fol. 16.] The number of riuulets becks & streames whose banks are beset with willowes & Alders wch giue occasion of easier fishing & slooping to the water makes that [bir crossed out] handsome coulered bird abound wch is calld

Alcedo Ispida or the King fisher. they bild in holes about grauell pitts [have their nests very full crossed out] wherein [are crossed out] is [above] to bee found great quantitie of small fish bones. & lay [a crossed out] very handsome round & as it were polished egges.

An Hobby bird[34] so calld becaus it comes in ether with or a litle before the Hobbies in the spring. of the bignesse of a Thrush coloured & paned[J] like an hawke marueliously subiet to the vertigo & and are sometimes taken in those fitts.

[34] This is evidently the Wryneck, which we now call the "Cuckoo's Mate," probably for the same reason that Browne associates it with the Hobby. It may be that the Hobby having become comparatively scarce, it was necessary to find another travelling companion for this bird, and that the Cuckoo was chosen as the most suitable. Old Norfolk names are Emmet-eater, and in one old book it is called Turkey-bird in a MS. note.

[J] That is marked with a barred or checkered pattern.

Upupa or Hoopebird[35] so named from its note a gallant marked bird wch I have often seen & tis not hard to shoote them.

[35] The Hoopoe would seem from this note to have been of more frequent occurrence than in the present day, see also in his answer to "Certain Queries" (Tract iv., Wilkin iv., p. 183), in which he says of this bird, "though it be not seen every day, yet we often meet with it in this country."

Ringlestones[36] a small [bird crossed out] white & black bird like a wagtayle & seemes to bee some kind of motacilla marina co[=m]on about yarmouth sands. they lay their egges in the sand & shingle about june and as the eryngo diggers tell mee not sett them flat butt upright likes [sic] egges in [a crossed out] salt.

[36] The Ring Plover is evidently the bird here referred to, but I have never

known the name of Ringlestone applied to this species in Norfolk, nor have I met with it elsewhere. The Eryngo is now no longer an article of commerce, and its diggers are extinct, but not their tradition as to the position in which the eggs of this bird are said to be placed--a "vulgar error" which does not accord with the writer's experience. When the full complement of four eggs is laid, they are arranged with their pointed ends towards the centre of the nest, which is a slight hollow in the soil. The concavity of the nest therefore, as well as the disproportionate size of the larger end, gives the eggs somewhat the appearance of being placed in the position referred to, but the small end of the egg is always visible, Sir Thomas Browne does not seem to have been aware of the remarkable fact of this essentially marine bird habitually nesting on the sandy warrens about Thetford in the south-west of Norfolk, far from the sea, which it still does, though in reduced numbers, and is there known as the Stone-hatch, from its habit of paving its nest with small stones.

The Arcuata or curlewe frequent about the sea coast.

[Fol. 17.] There is also an handsome tall bird Remarkably eyed and with a bill not aboue 2 inches long co[=m]only calld a stone curlewe[37] butt the note thereof more resembleth that of a green plouer [it crossed out] & breeds about Thetford about the stones & shingle of the Riuers.

[37] This characteristic Norfolk bird is still far from rare in the locality named by Browne, and is found in several other parts of the county. Willughby says, "The learned and famous Sir Thomas Brown, Physician in Norwich," informed him to the same effect, and repeats that its note (one of the most charming sounds uttered on the wild trackless heath on a summer's night) resembles that of the Green (i.e., Golden) Plover, but in the ear of the writer it is even more musical. In the third letter to Merrett, Browne says that he has kept the Stone Curlew (not "four Curlews," as Wilkin has it,) in large cages.

Auoseta[38] calld [I thinck a Barker crossed out] shoohingg-horne [written above] a tall black & white bird with a bill semicircularly reclining or bowed upward so that it is not easie to conceiue how it can feed answerable vnto the Auoseta Italorum in Aldrovandus a summer marsh bird & not unfrequent in Marshland.

[38] The Avoset is another bird which formerly frequented the marshy districts of Norfolk at the breeding time, but which has now been lost to us except as a very rare passing migrant in the spring. It probably ceased to breed in this county in or about the year 1818, and is said to have been exterminated in consequence of the demand for its feathers for the purpose of dressing artificial flies. It was called "Shoeing-horn," from the peculiar form of its beak, which, however, rather resembles the bent awl used by shoemakers. Girdlestone, who knew the bird well in its breeding haunts at Salthouse and Horsey, called it "Shoe-awl," a much more appropriate name. In his third letter to Merrett, Browne again mentions this bird, and applies to it the name of "Barker" (which he had crossed out in the above note), remarking that it was so called from its barking note. Jonston figures this bird twice; once in Tab. 48 under the name of Avosetta Italor., i.e., the Avosetta of the Italians, and again in Tab. 54 under the second name Avoselta species, an obvious error.

[A bird calld Barker from the note it hath crossed out]

A yarwhelp[39] so thought to bee named from its note a gray bird intermingled with some yellowish [whitish written above] fethers [the bill crossed out] somewhat long legged & the bill about an inch & half. esteemed a dayntie dish.

[39] This paragraph is written on the back of fol. 16. The Yarwhelp is the name by which the Black-tailed Godwit, a species which formerly nested in abundance in the marshes about Horsey and some adjacent localities in the Broads, was known. It virtually ceased to nest here sometime between the years 1829 and 1835, but perhaps an instance or two may have occurred rather later. It was also known as the "Shrieker." Browne again refers to this bird in the fourth letter to Merrett, where he calls it "barker" (a name which he had no doubt erroneously previously applied to the Avoset), or "Latrator, a marshbird,

about the bigness of a Godwitt," and once again under the name of "Yare-whelp, or barker," in his fifth letter; it may be that the name "barker" was applied indiscriminately to either species. As Lubbock names this bird as one of the "five species in particular" which "used formerly to swarm in our marshes" ("Fauna of Norfolk"), one would have thought Browne would have been better acquainted with it than seems to have been the case from the hesitating way in which he uses the vernacular name.

Loxias or curuirostra a bird a litle bigger than a Thrush of fine colours & prittie note [the m crossed out] differently from other birds, the [lower crossed out] upper & lower bill crossing each other. of a very tame nature, comes about the beginning of summer. I have known them kept in cages butt not to outliue the winter.

A kind of coccothraustes calld a [cobble crossed out] coble bird[40] bigger than a Thrush, finely coloured & shaped like a Bunting [it comes crossed out] it is [sometimes crossed out] chiefly [written above] seen [about crossed out] in su[=m]er about cherrie time.

[40] The Hawfinch was evidently not a very well-known bird in Browne's time, either to himself or Willughby; the latter says, "it is said to build in holes of trees." It has steadily increased in frequency as a breeding species with us for the last twenty years.

[fol. 16 verso.] A small bird of prey[41] [something smeared out here] calld a birdcatcher about the bignesse of a Thrush and linnet coloured with a longish white bill & sharpe of a very feirce & wild nature though kept in a cage & fed with flesh. [Added after in same hand but fresher ink] a kind of Lanius [Lanius crossed out and written more distinctly under].

[41] This paragraph is written on the back of fol. 16. The Red-backed Shrike, Lanius collurio, is the only species of Lanius mentioned by Browne; it is singular that he omits all mention of another bird, and that an essentially Norfolk species which would have been new to the Pinax--the Bearded

Titmouse, afterwards known to Edwards as the Least Butcher Bird. Browne certainly sent a drawing of this bird to Ray, who in his "Collection of English words not generally used" (1674), as pointed out by Mr. Gurney, mentions it as a "little Bird of a tawny colour on the back, and a blew head, yellow bill, black legs, shot in an Osiar yard, called by Sr. Tho. for distinction sake silerella," the drawing of which he acknowledges he had received. Pennant, 1768 ("Brit. Zool.," i., p. 165), follows Edwards ("Nat. Hist. of Birds," &c., 1745), who classes it with the Laniidae, and it was not till long after, and as the result of much discussion, that it was finally established as the only representative of a new genus under the name of Panurus biarmicus. The local name is Reed Pheasant, but Browne's name of Silerella seems an exceedingly appropriate one.

[p. 17 resumed.] A Dorhawke[42] or kind of Accipiter muscarius conceiued to haue its name from feeding upon flies & beetles. of a woodcock colour but paned like an Hawke a very litle poynted bill. large throat. breedeth with us & layes a maruellous handsome spotted egge. Though I haue opened many I could neuer find anything considerable in their mawes. caprimulgus.

[42] Browne seems to have been much interested in this remarkable bird, and mentions it again in his second and third letters to Merrett, especially in the latter; he calls it Caprimulgus, but conceives it to be a kind of Accipiter, muscarius, or cantharophagus, "in brief" [?] "avis rostratula gutturosa, quasi coxans, scarabaeis vescens, sub vesperam volans, ovum speciassisimum excludens," a fair specimen of the descriptive method of the time. Although he used the name Caprimulgus, it will be observed that he does not mention the "vulgar error" which led to its being so called. Merrett includes this species in the Pinax under the name of "Caprimulgus, or the Goat-sucker," but in a letter to Browne tells him he knows no Hawk called a Dorhawk.

[Fol. 18.] Auis Trogloditica[43] or Chock a small bird mixed of black & white & breeding in cony borrouges whereof the warrens are full from April to September. at which time they leaue the country. they are taken with an Hobby and a net and are a very good dish.

[43] The Wheatear is here referred to; the name trogloditica would seem to be more appropriate in this country, having reference to its habits of nesting in "Cony borroughs," than that of aenanthe, as applied to it by those who knew it as frequenting the Continental vineyards. A name still, or recently in use in West Norfolk, is Cony-chuck.

Spermologus. [sic] Rookes wch by reason of the [in reason of crossed out] great quantitie of corn feilds & Rooke groues are in great plentie the yong ones are co[=m]only eaten sometimes sold in norwich market & many are killd for their Liuers in order to cure of the Rickets.

Crowes[44] as euerywhere and also the coruus variegatus or pyed crowe with dunne & black interchangeably they come in the winter & depart in the summer & seeme to bee the same wch clusius discribeth in the faro Islands from whence perhaps these come. [they are crossed out] and I have seen them [written above] very co[=m]on in Ireland, butt not known in many parts of England.

[44] The Crow (Corvus corone) is much less common in Norfolk than formerly, but it still nests here in a few scattered localities. C. cornix, the Hooded, Norway, Danish, or "Royston" Crow, is an autumn immigrant as of yore, but not especially from the Faroee Islands; both species (or forms as by some regarded) are immigrants from the east, but the latter, as a rule, occupies a more northern range than the former. The Raven (C. corax) is now a very rare visitor to Norfolk; it is probable that it last nested in this county in the year 1859. The Jackdaw, or Caddow, is common enough, but the Chough (Pyrrhocorax graculus) is quite unknown in Norfolk. Although the Magpie must have been well known to Browne I find no mention of it in these notes.

Coruus maior Rauens in good plentie about the citty wch makes so few Kites to bee seen hereabout. they build in woods very early & lay egges in februarie.

Among the many monedulas or Jackdawes I could neuer in these parts obserue the pyrrhocorax or cornish chough with red leggs & bill to bee co[=m]only seen in Cornwall. & though there bee heere very great [num crossed out] store of partridges yet [not crossed out] the french Red leggd partridge[45] is not to bee met with [heere crossed out]. the Ralla or Rayle[46] wee haue counted a dayntie dish. as also no small number of Quayles. the Heathpoult[47] co[=m]on in the north is vnknown heere as also the Grous. though I haue heard some haue been seen about Lynne. the calandrier or great [Fol. 19] great [sic] crested lark Galerita I haue not met with heere though with 3 other sorts [of Larkes written above] the ground lark woodlark & titlark.

[45] The Red-legged Partridge is now common enough; it was introduced into the Eastern Counties at Sudbourne and Rendlesham, in East Suffolk, in or about the year 1770, by both the Marquis of Hertford and Lord Rendlesham. How quickly they established themselves may be judged from the fact that in the season of 1806-7 of 1,927 Partridges killed at Rendlesham 112 were Red legs, but they do not seem to have spread very far. A second introduction, this time into West Suffolk, much nearer to the Norfolk border, at and about Culford, was effected in the year 1823, and from this centre they rapidly spread into Norfolk, in which county also others were imported by the resident proprietors.

[46] The Land Rail (Crex pratensis) or Daker hen, is doubtless here referred to, as the Water Rail has already been mentioned (p. 15 ante) as "a kind of Ralla aquatica." This bird is a summer visitor, by no means common and very uncertain in its numbers. The same applies to the Quail, which appears to be less frequent than formerly, no doubt from the great destruction on the Mediterranean coast in spring of the birds migrating to England. In the summer and autumn of 1870 we had an unusual influx of these latter birds.

[47] How far the indigenous race of Blackgame, which undoubtedly lingered for many years about Wolferton and Sandringham, still exists, it is difficult to say; examples turn up occasionally, but so many of these birds have been introduced and turned off in different parts of the county in the course of the past forty years, that it is impossible to speak with certainty.

Stares or starlings in great numbers. most remarkable in their [great crossed out] numerous [written above] flocks [about the crossed out] wch I haue obserued about the Autumne when they roost at night [up crossed out] in the marshes in safe place upon reeds & alders. wch to obserue I went to the marshes about sunne set. where standing by their vsuall place of resort I obserued very many flocks flying from all quarters. wch in lesse than an howers space came all in & settled in innumerable [quantitie crossed out] numbers [written below] in a small compasse.

Great varietie of finches[48] and other small birds whereof one very small [one crossed out] calld a whinne bird marked with fine yellow spotts & lesser than a wren. there is also a small bird called a chipper somewhat resembling the former wch comes in the spring & feeds upon the first buddings of birches & other early trees.

[48] In his fifth letter to Merrett Browne says, "I confess for such little birds I am much unsatisfied on the names given to many by countrymen and uncertain what to give them myself." This is painfully apparent in the cases of the two little birds here referred to as the "Whinne-bird" and the "Chipper." From the description of the former, "marked with fine yellow spots and lesser than a Wren," also with a "shining yellow spot on the back of the head," it seems likely that the Gold-crested Wren is intended. The Chipper, he says, "comes in the spring and feeds upon the first buddings of birches and other early trees;" he also calls it "Betulae carptor," and says that he sends a drawing to Merrett; a third mention is as follows: "That which I called a Betulae carptor, and should rather have called it Alni carptor ... it feeds upon alder buds, nucaments, or seeds, which grow plentifully here; they fly in little flocks." I can only suggest that this bird may be the Siskin, which fairly answers the description. It visits us in small flocks on its way north very early in the year, feeding upon the seeds of the alder, birch, and larch trees. One would however have thought that the Siskin would have been well known to Browne, as it evidently was to Turner, Willughby, and Ray. Merrett mentions it under Turner's name of "Luteola."

A kind of Anthus [or crossed out] Goldfinch [written above] or fooles coat co[=m]only calld a drawe water. finely marked with red & yellowe & a white bill. wch they take with trap cages in norwich gardens & fastning a chaine about them tyed to a box of water it makes a shift with bill and legge to draw up the water unto it from the litle pot hanging [abot the length of crossed out] by the chaine about a foote [downe crossed out] belowe.

[The account of the Roller, which is written on smaller paper, will be found improperly inserted among the Fishes, between pp. 30 and 32 as follows:--]

[Fol. 31.] On the xiiii of May 1664 a very rare bird was sent mee kild about crostwick wch seemed to bee some kind of Jay.[49] the bill was black strong and bigger then a Jayes somewhat yellowe clawes tippd black. 3 before and one clawe behind the whole bird not so bigge as a Jaye [the crossed out.]

[49] This note is interesting as the first record of the occurrence of the Roller in Britain, to which country it is a rare wanderer. Although it had long been known on the Continent, its identity seems to have puzzled Browne, and he imagines (as did others, both before and after him,) it to be some kind of Jay; later, in his second letter to Merrett (January, 1668), he says that it answers to the description of Garrulus argentoratensis (the name given by Aldrovandus to whom it was known), and calls it "the Parrot-jay." This is five years after the original note was made, and we find that the words Garrulus argentoratensis, written by the same hand but with a different pen and ink, have been added subsequently, doubtless as the result of further information. In another letter he mentions having sent the bird to Merrett, but adds, "If you have it before I should bee content to have it againe otherwise you may please keep it."

The head neck & throat of a violet colour the back upper parts of the wing of a russet yellowe the fore & part of the wing azure succeeded downward by a greenish blewe then on the flying feathers bright blewe the lower parts of the wing outwardly of a browne [the crossed out] inwardly of a merry blewe the belly a light faynt blewe the back toward the tayle of a purple blewe the tayle eleuen fethers of a greenish coulour the extremities of the outward fethers thereof white wth an eye[K] of greene. Garrulus Argentoratensis [the name added in a different ink and pen].

[K] Tinge, shade, particularly a slight tint.--"Imp. Dict."

NOTES ON CERTAIN FISHES AND MARINE ANIMALS FOUND IN NORFOLK.

[MS. SLOAN. 1882. FOL. 145-146. ALTERED TO 21 AND 22, AND 1830 FOL. 23-30 AND 32-38.]

[The introductory remarks, paragraphs one to three, will be found in the volume of the Sloane MSS. numbered 1882 (labelled "Notes on Generation"), on pages 145 and 146, which are altered to 21 and 22. They were placed in their present position by Wilkin, but although appropriate, there is nothing to show that they belong to the set of notes here reproduced, and they may form memoranda for the beginning of some essay never completed. The contents of the volume in question are of a very miscellaneous character, and consist of fragmentary notes, which appear to be memoranda jotted down at random.]

[Fol. 21/145.] It may well seeme no easie matter to giue any considerable account of fishes and animals of the sea wherein tis sayd that there are things creeping innumerable both small and great beasts because they liue in an element wherein they are not so easely discouerable notwithstanding probable it is that after this long nauigation search of the ocean bayes creeks Estuaries and riuers that there is scarce any fish butt hath been seen by some man for the large & breathing sort thereof do sometimes discouer themselues aboue water and the other are in such numbers that some at one time or other they are discouered and taken euen the most barbarous nations being much addicted to fishing and in America and the new discouered world the people were well acquantd with fishes of sea and riuers, and the fishes thereof haue been since described by industrious writers.

Pliny seemes to short in the estimate of their number in the ocean, who recons up butt one hundred & seventie six species. butt the seas being now farther known & searched [21/145 verso] Bellonius much enlargeth.

and in his booke of Birds thus deliuereth himself allthough I think it impossible to reduce the same vnto a certain number yet I may freelie say that tis beyond the power of man to find out more than fiue hundred sorts [kinds written above] of fishes three hundred sorts of birds more than three hundred sorts of fourfoted animalls and fortie diversities of serpents.[50]

[50] This estimate of the number of species of birds and fishes existing is amusing in the light of the present knowledge of the subject. Of course any such estimate can only be approximate, and open to constant emendation; but according to a statement in the "Zoological Record" of 1896, it was believed that there were something like 386,000 described species: 2,500 of which are mammals, 12,500 birds, 4,400 reptilia and batrachia, 12,000 fishes, 50,000 mollusca, 20,000 crustacea, and 250,000 insecta; the smaller divisions I have omitted. And whereas only about 10,000 species of plants were known to Linnaeus, Professor Vines in his address to the Botanical section at the Bradford meeting of the British Association, 1900, states that the approximate number of recognised plants at present existing is 175,596; but this is far short of the total of existing species. Professor Saccardo states that there are 250,000 fungi alone, and that the number of existing species in other groups would bring the total up to over 400,000.

[SLOANE MSS. 1830, FOL. 23-38.]

[Fol. 23.] Of fishes sometimes the larger sort are taken or come ashoar. A spermaceti whale[51] of 62 foote long neere Welles. another of the same kind 20 yeares before at Hunstanton. & not farre of 8 or nine came ashoare & 2 had yong ones after they were forsaken by ye water.

[51] In the muniment room at Hunstanton Hall there exists a book of MSS. notes relating to their estates, kept by Sir Hamon and Sir Nicholas le Strange, between the years 1612 and 1723. From this book Mr. Hamon le Strange has been good enough to send me an extract containing the full particulars of the stranding and disposal of a Sperm Whale 57 feet long, which came ashore on their Manor of Holme, on the 6th December, 1626, the skull of which is still in the courtyard at Hunstanton Hall.

Browne had not come to reside in Norwich at that time, and the chapter on the Spermaceti Whale in his Pseudodoxia Epidemica, was inspired by a subsequent occurrence of the same kind, for, as appears from the above note, a larger individual, 62 feet long, came ashore at Wells 20 years later, which he says led him to further inquiry. This would indicate about the year 1646 as the date of the latter occurrence, whereas in his third letter to Merrett, written in 1668, he states that it happened "about 12 years ago," or in 1656. There is probably an error in one of these dates.

Another example seems to have been found at Yarmouth about the year 1652, for we find Browne writing in that year for particulars of its "cutting up." (See Appendix E.)

In the postscript to a letter also in the muniment room at Hunstanton, dated June 11th, 1653, written to Sir Hamon le Strange, who had been consulting him professionally, Browne says: "I pray you at your leisure doe mee the honor to informe mee how long agoe the Spermaceti Whale was cast upon your shoare & whether you had any spermm with in any other part butt the head." It will be noticed that in both the letters referred to he is anxious to ascertain in what part of the body the "sperm" was situated, doubtless for the purpose of confuting the "vulgar conceit" as to the origin of the "sperm" referred to in the second paragraph of his treatise in the Pseudodoxia. His investigations also probably first led to a certain knowledge as to the nature of the food of this animal.

These, however, although the first to be recorded in this county, were not the first or only occurrences of the kind, for there is in the parish church of Great Yarmouth the base of the skull of a Sperm Whale, used as a chair, for the painting of which a charge of five shillings appears in the churchwardens' accounts for the year 1606; many such events in European waters are to be found recorded.

But the most interesting circumstance with regard to these whales is the statement that "two had yong ones after they were forsaken by the water." This event renders it highly improbable that they were Sperm Whales, for the stragglers of that species which have been met with in our waters, and indeed in the northern seas generally, have been almost invariably solitary males, or, in one or two instances "schools" of young males. In the only instance in which both sexes were found, the school was composed I believe of immature individuals. (Vide J. Anderson, "Nachrichten von Island, Groenland, und der Strasse Davis," Frantfurt (1747), p. 248.) Moreover, this view is confirmed by a letter which will be found in Appendix B., where the following passage occurs:--"And not only whales, but grampusses have been taken in this Estuarie ... and about twenty years ago four were run ashore near Hunstanton, and two had young ones after they had come to land." A so-called Grampus which came ashore on the 21st July, 1700, was from a description and drawing in the le Strange MS. above quoted, a male Hyperoodon rostratus, apparently nearly adult.

The Grampus (Orca gladiator) (mentioned in the next paragraph) is frequently met with in the British seas, and has repeatedly occurred on the Norfolk coast. Some early occurrences are on record, for instance in Mackerell's "History of Lynn," twelve are said to have come ashore near that town in 1636, and another in 1680. Two very juvenile examples were taken off Yarmouth in November 1894.

A grampus aboue 16 foot long taken at yarmouth [3 or crossed out] 4 yeares agoe.

The Tursio or porpose is co[=m]on the Dolphin[52] more rare though sometimes taken wch many confound with the porpose. butt it hath a more waued line along the skinne sharper toward ye tayle the head longer and nose more extended wch maketh good the figure of Rondeletius. the flesh more red & [fa crossed out] well cooked of very good taste to most palates & exceedeth that of porpose.

[52] There can be no doubt that the Common Dolphin (Delphinus delphis) is here referred to, and indeed this species might reasonably be expected to be met with on our coast, as its range extends at least as far to the north as the Scandinavian waters, but so far as the writer is aware Browne's is the only record of its having been met with in Norfolk. The White-beaked Dolphin (D. albirostris) is not unfrequent, but it is clear that Browne does not refer to that species.

In the "Vulgar Errors," Browne devotes a whole chapter (chapter ii. of the fifth book) to a learned treatise on the "Picture of Dolphins," and in one of the letters to his son Edward (Sloane MSS., 1847), dated June 14th [1676?], he writes feelingly as an anatomist, evidently fearing that a specimen then available might be wasted, instead of being reserved for scientific purposes; for, says he, "if the dolphin were to be showed for money in Norwich, little would bee got; if they showed it in London they are like to take out the viscera, and salt the fish, and then the dissection will be unconsiderable." He then refers to the dolphin "opened when the King was here," and describes its anatomical peculiarities, adding that Dame Browne cooked the flesh "so as to make an excellent savory dish of it," and that "collars" thereof (steaks cut transversely) being sent to the King, who was then at Newmarket, for his table, they "were well liked of." It is evident therefore that he was present at the dissection of two of these animals.

The vitulus marinus[53] seacalf or seale wch is often taken sleeping on the shoare [4 crossed out] 5 [written above] yeares agoe one was shot in the riuer of norwich about surlingham [wh crossed out] ferry having continued in the riuer for diuers moneths before being an Amphibious animal it may bee caryed about aliue & kept long if it can bee brought to feed some haue been kept many moneths in ponds. the pizzell the bladder the cartilago ensiformis the figure of the Throttle the clusterd & racemous forme of the kidneys [Fol. 24] the flat & compressed heart are remarkable in it. in stomaks of all that I have opened I have found many [short crossed out] wormes.

[53] There is in the present day a considerable number of Common Seals inhabiting the sand-banks of the Wash between the Norfolk and Lincolnshire coasts, and they are frequently captured by the fishermen; nor has the habit of straying into fresh-water deserted them, for in recent years they have been taken in the River Ouse at Bluntisham, forty miles from the sea. Three other species of Seal have been taken on the Norfolk coast, viz., Phoca hispida, P. barbata, and Halichoerus gryphus.

I haue also obserued a scolopendra cetacea[54] of about ten foot long answering to the figure in Rondeletius wch the mariners told me was taken in these seas.

[54] A Scolopendra, ten feet long, is at first rather startling, but on referring to Rondeletius's Libri de piscibus Marinis (lib. xvi. p. 488), I find that under the name "Scolopendra" he includes at least three distinct forms--i., S. terrestris, a centipede; ii., S. marina, certain species of Nereidiform polychaet worms; iii., Scolopendra cetacea, regarded as a Cetacean and figured with a Cetacean blow-hole. With regard to this remarkable figure my friend, Dr. S. F. Harmer, has favoured me with the following note:--"In the account given Rondeletius is evidently writing from report; the figure is also no doubt borrowed, and may have been 'improved' when redrawn; it seems to me that it is based upon some kind of Tunny, although he figures a Tunny earlier in the book (lib. viii. p. 249). The idea of the lateral appendages might have been derived from the dorsal and ventral finlets of a Tunny; but the first four finlets on each side are imaginary structures, and in a wrong position. I can offer no opinion with regard to the nasal appendages." Jonston (De piscibus, p. 156, Tab. xliv.) also gives a similar figure of Scolopendra Cetacea, which appears to be a further modification of Rondeletius's figure; here it has teeth, shown like those of the Sperm Whale, and an extra dorsal-fin is added; the number of

lateral appendages is the same, and a column of water proceeding from the blow-hole is falling gracefully forward. It is worthy of notice that Rondeletius also figures the Saw-fish [Pristis] with a blow-hole.

A pristes or serra [written above] saw fish[55] taken about Lynne co[=m]only mistaken for a [sha crossed out] sword fish & answers the figure in Rondeletius.

[55] In the "Transactions of the Linnean Society," ii., p. 273, is an essay by Latham "On the various species of Sawfish," but he does not mention any British locality. So far as I am aware Browne's is the only record of the occurrence of this southern species in British waters, with the exception of a note in Fleming's "British Animals," p. 164, where it is stated on the authority of the late Dr. Walker's MS. "Adversaria" for 1769, that Pristis antiquorum is "found sometimes in Loch Long," but Fleming adds that he has met with no other proof of its ever having visited the British shores. Browne mentions in his eighth letter to Merrett that he sends him a "figure in little" of a Pristis which he received of a Yarmouth seaman, and is so precise in his statement that his fish was Pristis serra (the Pristis antiquorum of Cuvier), that his record cannot be disregarded. He specially guards against its being mistaken for the Sword-fish (Xiphias gladius), which has been taken on several occasions in our waters, and of which he gives some interesting particulars.

A sword fish or Xiphias or Gladius intangled in the Herring netts at yarmouth agreable unto the Icon in Johnstonus with a smooth sword not vnlike the Gladius of Rondeletius about a yard & half long, no teeth [n crossed out] eyes very remarkable enclosed in an hard cartilaginous couercle about ye bignesse of a good apple. ye vitreous humor plentifull the crystalline larger then a nutmegge [cleare crossed out] remaining cleare sweet & vntainted when the rest of the eye was vnder a deepe corruption wch wee kept clear & limpid many moneths vntill an hard frost split it & manifested the foliations thereof.

It is not vnusuall to take seuerall sorts of canis or doggefishes[56] great and small wch pursue the shoale of herrings and other fish butt this yeare 1662 one was taken intangled in the Herring netts about 9 foot in length, answering the last figure of Johnstonus lib 7 vnder the name of canis carcherias alter & was by the teeth & 5 gills one kind of shark particularly [Fol. 25] remarkable in the vastnesse of the optick nerves & 3 conicall hard pillars wch supported the extraordinarie elevated nose wch wee haue reserued with the scull the seamen calld this kind a scrape.

[56] Various species of Dog-fish are frequent off the Norfolk coast as elsewhere. The name "Sweet William" is applied to the larger fish of this kind, especially to the Tope; this appears also to have been the case in Pennant's time, for alluding to this vernacular name he supposes it was applied in ironical allusion to the offensive smell of their flesh and skin. They are objects of great aversion among the fishermen, owing to the disturbance they create among the shoals of fish, and the damage they do to both nets and the enclosed fish. Scarcely a season passes but one or more specimens of Browne's Canis carcharias, or, as modern Ichthyologists call it, Lamna cornubica, the Porbeagle, being entangled in the drift nets and landed with the herrings. One lies on the fish-wharf at Lowestoft as I write this note on the 19th of October, 1900, measuring 7 feet 10 inches in length. Jonston's figure referred to by Browne is evidently intended for this species, but he makes a slight error in the reference to the Historia Naturalis (De Piscibus et Cetis); it occurs in book v., and the figure is fig. 6 on Tab. vi., and it is marked Canis carcharias alius (not alter).

Sturio or Sturgeon[57] so co[=m]on on the other side of the sea about the mouth of the elbe come seldome into our creekes though some haue been taken at yarmouth & more in the great [owse crossed out] Owse by Lynne butt their heads not so sharpe as represented in the Icons of Rondeletius & Johnstonus.

[57] So great is the variation in the snout of the Sturgeon, that Dr. Parnell in his excellent essay on "The Fishes of the District of the Forth," describes the Sharp-nosed Sturgeon as a distinct species under the name of Acipenser sturio, and the broad-nosed form he calls A. latirostris. His views, however, have not

been generally accepted, and only one British species is recognised. The Sharp-nosed variety has been taken here, but the normal form is much more frequent.

Sometimes wee meet with a mola or moonefish[58] so called from some resemblance it hath [from crossed out] of a crescent in the extreme part of the body from one finne unto another one being taken neere the shoare at yarmouth before breake of day seemed to shiuer & grunt like an hogge as Authors deliuer of it the flesh being hard & neruous it is not like to afford a good dish butt from the Liuer wch is [white crossed out] large white & tender somewhat [wee crossed out] may bee expected [for crossed out] the gills of these fishes wee found thick beset with a kind of sealowse. [Added subsequently] in the yeare 1667 a mola was taken at monsley wch weighed 2 [p crossed out] hundred pound.

[58] This fish (Orthagoriscus mola), which we know as the Sun-fish, has been repeatedly taken here. For an account of its parasites see Cobbold on the "Sun-fish as a host," "Intellectual Observer," ii., p. 82; also Day, "Brit. Fishes," ii., p. 275. According to Dr. Spencer Cobbold the Sun-fish is infested by nine species of Helminths, three of which are mostly found attached to the gills, while a fourth adheres to the surface of the body.

The Rana piscatrix or frogge fish[59] is sometimes found in a very large magnitude & wee haue taken the [paynes crossed out] care [written above] to haue them clend & stuffed. wherein wee obserued all the appendices whereby the[y] cach fishes butt much larger then are discribed in the Icons of Johnstonus tab xi fig 8.

[59] Both this species and the Wolf-fish are well known upon our coast.

[Fol. 26] The sea [wollf crossed out] wolf or Lupus nostras of Schoneueldus remarkable for its spotted skinne & notable teeth incisors Dogteeth & grinders the dogteeth [in the crossed out] both in the jawes & palate scarce answerable by any fish of that bulk for [strength crossed out] the like disposure strength &

soliditie.

Mustela marina[60] called by some a wesell ling wch salted & dryed becomes a good Lenten dish.

[60] Some member of the family Gadidae is here referred to, probably the five-bearded Rockling, Motella mustela, or Brown Whistle-fish of Pennant, which is occasionally taken by our fishermen, but is by no means common.

A Lump or Lumpus Anglorum so named by Aldrouandus by some esteemed a festiuall dish though it affordeth butt a glutinous jellie & the skinne is beset with stony knobs after no certaine order ours most answereth the first figure in the xiii table of Johnstonus butt seemes more round & arcuated then that figure makes it.

Before the herrings there co[=m]only cometh a fish about a foot long by the fish man called an horse[61] resembling in all poynts the Trachurus of Rondeletius of a mixed shape between a mackerell & an herring. obseruable from [an oblique bo crossed out] its greene eyes rarely skye colored back after it is kept a day & an oblique bony line running on ye outside from the gills vnto ye tayle. a drye & hard dish butt makes an handsome picture.

[61] This is the Horse Mackerel, or Scad, Caranx trachurus; a handsome fish and common enough, especially off Sheringham, but not much esteemed for the table.

The Rubelliones or Rochets[62] butt thinly met with on this coast. the gornart cuculus or Lyrae species more often wch they seldome eat butt bending the back & sprdding the finnes into a liuely posture do hang them up in their howses.

[62] Fish of the Gurnard kind are here referred to. The Rochet of Pennant is the Red Gurnard, Trigla cuculus; he calls T. lyra the Piper. Large numbers of various species of Gurnard are brought in by our trawlers and sell readily,

especially the Sapphirine Gurnard, or Tub-fish (T. hirundo), which is known as the "Lachet" on our coast; it reaches a large size, and seems to be much in demand for the table. In spring the colours are very brilliant, and they are frequently seen on the fish stalls with their pectoral fins extended as Browne describes.

[Fol. 27.] Beside the co[=m]on mullus[63] or mullet there is another not vnfrequent wch some call a cunny fish butt rather a red muellett of a flosculous redde & somewhat rough on the scales answering the discription of [Rond crossed out] Icon of Rondeletius vnder the name of mullus ruber asper [no crossed out] butt not the tast of the vsually knowne mullet as [being butt crossed out] affording butt a drye & leane bitt.

[63] The Common Mullet I take to be the Grey Mullet (Mugil capito), which is at times plentiful on our coast, coming into Breydon and the mouths of the rivers, but the Red Mullet (Mullus barbatus) is far less frequently met with. In his third letter to Merrett, Browne says, "There is of them maior and minor," the latter probably being the variety known as the Surmullet, by far the most frequently met with here.

Seuerall sorts of fishes[64] there are wch [bear crossed out] do [written above] or may beare the names of seawoodcocks as the Acus maior scolopax & saurus. the saurus wee sometimes meet with yonge. Rondeletius confesseth it a very rare fish somewhat resembling the Acus or needlefish before & a makerell behind. wee have kept one dryed many yeares agoe.

[64] The Saurus of Rondeletius appears to be the Skipper or Saury-pike (Scombresox saurus) of modern authors. Acus major is the Gar-fish or Greenback (Belone vulgaris); this is the Acus primus of Rondeletius, Dr. Harmer has been good enough to send me the following note on Rondeletius's figures:--"De Acus secunda specie" (lib. viii. p. 229). "Two species are figured; the upper figure appears to represent Siphonostoma typhle, and the lower one S. acus. Guenther ('Brit. Mus. Cat.,' viii. p. 157) gives a reference to Rondeletius in his synonyms of S. acer without indicating that the latter

figures two species. Under S. typhle (p. 154) he gives the synonym Syngnathus rondeletii, De la Roche. A reference to Delaroche ('Ann. Mus. Hist. Nat., Paris,' xiii, 1809 p. 324, Pl. xxi. fig. 5) shows that S. rondeletii is identified with the first figure on p. 229 of Rondeletius; and it may thus be concluded that Guenther agrees with this conclusion. It seems therefore probable that Browne's Acus of Aristotle refers to S. typhle."

The Acus maior calld by some a garfish & greenback answering ye figure of Rondeletius under the name of Acus prima species remarkable for its quadrangular figure and verdigreece green back bone.

[L] A lesser sort of Acus [wee crossed out] maior or primae specaeei wee meet with [answering the saurus of Rondeletius crossed out] much shorter then the co[=m]on garfish & in taking out the spine wee found it not green as in the greater & much answering the saurus of Rondeletius.

[L] This and the next paragraph on the back of Fol. 26 are in different ink and smaller writing though in the same hand, and appear to have been added subsequently. The first paragraph is omitted by Wilkin.

A scolopax[65] or sea woodcock of Rondeletius was giuen mee by a seaman of these seas. about 3 inches long & seemes to bee one kind of Acus or needlefish answering the discription of Rondeletius.

[65] The Scolopax, or Sea Woodcock, is clearly Centriscus scolopax, a very rare fish in the British seas, and it would have been well had Browne given a more precise account of the origin of his specimen.

The Acus of Aristotle [see Note 64] lesser thinner corticated & sexangular by diuers calld an addercock & somewhat resembling a snake ours more plainly finned then Rondeletius discribeth it.

A little corticated fish[66] about [4 inches crossed out] 3 or 4 inches long [several words smeared out] ours answering that wch is named piscis octangularis by wormius, cataphractus by Schoneueldeus. octagonius versus caput, versus caudam hexagonius.

[66] Doubtless the Armed Bull-head, or Pogge, Agonus cataphractus. A MS. note in Berkenhout says it was called at Lowestoft a Beetle-head (1769).

[Fol. 28.] The faber marinus[67] sometimes found very large answering ye figure of Rondeletius. which though hee mentioneth as a rare fish & to be found in the Atlantick & Gaditane ocean yet wee often meet with it in these seas co[=m]only calld a peterfish hauing [a crossed out] one [written above] black spot on ether side the body conceued the perpetuall signature from the impression of St Peters fingers or to resemble the 2 peeces of money wch St Peter tooke out of this fish remarkable also from its disproportionable mouth & many hard prickles about other parts.

[67] Zeus faber, the Dory. Many, usually small ones, are brought in by our fishermen.

A kind of scorpius marinus[68] a rough prickly & monstrous headed fish 6 8 or 12 inches long answerable vnto the figure of Schoneueldeus.

[68] Cottus scorpius, Father Lasher, commonly taken by the shrimpers.

A sting fish[69] wiuer or kind of ophidion or Araneus slender, narrowe headed about 4 inches long wth a sharpe small prickly finne along the back which often venemously pricketh the hands of fishermen.

[69] Probably from its size the Lesser Weever, Trachinus vipera, as also the Draco minor of Jonstoni. A common fish in our waters. Large numbers of the Greater Weever, T. draco, are brought in by the trawlers.

Aphia cobites marina[70] or sea Loche.

[70] One of the Gobies. Day, "Brit. Fishes," i., p. 169, supposes the Aphya

cobites of Rondeletius (p. 20) to be the White Goby, A. pellucida; Pennant has A. cobites as a synonym for the Spotted Goby (G. minutus) and the Sea Gudgeons, Black Gobies (G. niger), but at that time there was no very nice distinction of the members of this genus. The Sea Miller's Thumb is probably the Shanny (Blennius pholis). Alosa, is the Allis Shad (Culpea alosa, L.), not uncommon (see Note 74).

Blennus a sea millars thumb.

Funduli marini sea gogions.

Alosae or chads to bee met with about Lynne.

Spinachus or smelt[71] in greatest plentie about Lynne butt [co[=m]on on yarmouth coast crossed out] where they haue also a small fish calld a primme answering in [all crossed out] tast & shape a smelt & perhaps are butt the yonger sort thereof.

[71] The Smelt, Osmerus eperlanus, is abundant in the shallow waters and estuaries on the Norfolk coast in spring, ascending the fresh-water rivers to spawn. The small fish called a Primme by Browne, may be the Atherine (Atherina presbyter), which is also found in our waters, where it is often mistaken for the Smelt, but I have not heard it called by the former name.

[Fol 29.] Aselli or cods of seuerall sorts. Asellus albus or whitings in great plentie. Asellus niger carbonarius or [col crossed out] coale fish. Asellus minor Schoneueldei callarias pliny or Haydocks with many more also a weed fish somewhat like an haydock butt larger & dryer meat. A Basse also much resembling a flatter kind of Cod.[72]

[72] The first three fishes named in this paragraph need no comment; the Weed-fish is doubtless a local name, but for what species I cannot discover. The Bass, Labrax lupus (Cuv.), is, as might be expected from the nature of our coast, by no means common here.

Scombri are makerells[73] in greate plentie a dish much desired butt if as Rondeletius affirmeth they feed upon sea starres & squalders (see Note 90) there may bee some doubt whether their flesh bee without some ill qualitie sometimes they are of a very large size & one was taken this yeare 1668 wch was by measure an ell long and of ye length of a good salmon, at Lestoffe.

[73] The latter part of this paragraph, beginning, "Sometimes they are of a very large size," is written on the left-hand side of the opening, and is evidently a subsequent addition. One would be inclined to think from the great size of the fish here recorded (3 ft. 9 in.), that it may have been a species of Tunny, or even a Bonito, both of which have been taken on the Norfolk coast. Seventeen inches is a large mackerel.

Herrings departed sprats or sardae not long after succeed in great plentie wch are taken with smaller nets [& dryed crossed out] & smoakd & dryed like herrings become a [daint crossed out] sapid bitt & vendible abroad.

Among these are found Bleakes or bliccae[74] a thinne herring like fishe wch some will also think to bee young herrings. And though the sea aboundeth not with pilchards, yet they are co[=m]only taken among herrings. butt few esteeme thereof or eat them.

[74] It is quite evident that the fish referred to here, and again in the sixth letter to Merrett, is not the true Bleak (Alburnus lucideus) of our freshwaters. It seems that the young of some species of Clupeoid was thus known, for I find it stated in a MS. note in a copy of Berkenhout's "Outlines of the Natural History of Great Britain," (1769), in the possession of Mr. T. E. Gunn, that the Bleak and the Sprat are often caught together in the sea at Aldeburgh (Suffolk) in November, and the writer of the note adds, "the Bleak is larger than the Sprat, its eyes are larger, and the upper part of its belly serrated." I think from this description and from Browne's remarks, that the young of a species of Shad must have been mistaken for the Bleak, which although found low down in our rivers almost to where the salt tide mingles with the fresh, does not I

believe enter the salt water.

Congers are not so co[=m]on on these coasts as on many seas about England, butt are often found upon the north coast of Norfolk, & in frostie wether left in pulks & plashes upon the ebbe of the sea.

[Fol. 30.] The sand eels Anglorum of Aldrouandus, or Tobianus of Schoneueldeus co[=m]only called smoulds taken out of the sea sands with forks & rakes about Blakeney and Burnham a small round slender fish about 3 or 4 inches long as bigge as a small Tobacco pipe a very dayntie dish.

Pungitius marinus[75] or sea bansticle hauing a prickle one each side the smallest fish of the sea about an inch long sometimes drawne ashoare with netts together with weeds & pargaments[M] of the sea.

[75] The smallest of the genus Gasterosteus, or Stanstickles, is G. pungitius, the ten-spined Stickleback, but this fish is two inches long when full grown. All the species seem to be more or less indifferent to the salinity of the water. The fifteen-spined Stickleback, G. spinachia, is also sometimes taken by the shrimpers, and is the most truly marine species, but is by no means "the smallest fish of the sea."

[M] This word which Wilkin renders "fragments," is doubtless from the Latin pergamentum, and it seems likely that Browne had in view certain sea-weeds, possibly Laminaria or Ulva which, especially when dry, present somewhat the appearance and texture of parchment.

Many sorts of flat fishes[76] The pastinaca oxyrinchus with a long & strong aculeus in the tayle conceuud of speciall venome & virtues.

[76] Pastinaca oxyrinchus appears to be the Sting Ray (Trygon pastinaca); Raia clavata, the Thornback; R. oculata, the Spotted Ray (R. maculata); R. aspera; the Shagreen Ray? (R. fullonica).

Severall sorts of Raia's skates & Thornebacks the Raia clauata oxyrinchus, raia oculata, aspera, spinosa fullonica.

The great Rhombus or Turbot aculeatus & leuis.

The passer or place.

Butts of various kinds.

The passer squamosus Bret Bretcock[77] & skulls comparable in taste and delicacy vnto the soale.

[77] The Brill, Rhombus laevis (Lin.), Passer asper squamosus_, Rondl., formerly known as the Brett, Bretcock, Skull, or Pearl.

The Buglossus solea or soale[78] plana & oculata as also the Lingula or small soale all in very great plentie.

[78] Solea vulgaris, the Common Sole. The "Lingula, or small Sole," is probably the Solea variegata, Flem., the S. parva sive Lingula of Rond. Jonston figures "Solea lingulata," Tab. xx., fig. 12, but I am uncertain what species is intended. It is possible that Browne may have Latinised the trade name by which small Soles are known in the market as "slips" and "tongues." What other species he may have wished to indicate as "plana" and "oculata" it is difficult to determine.

Sometimes a fish aboue half a yard long like a butt[79] or soale called asprage wch I haue known taken about Cromer.

[79] The "asprage" (or it may be "a sprage") may possibly be the Dab, Pleuranectes limanda, which Rondeletius calls Passer asper. I do not find that species mentioned otherwise, and a great many are taken by the Cromer and Sheringham fishermen.

[Fol. 31.] [See Roller ante p. 30.]

[Fol. 32.] Sepia or cuttle fish[80] [smear] & great plentie of the bone or shellie substance which sustaineth the whole bulk of that soft fishe found co[=m]only on the shoare.

[80] Of the various species of the Cephalopoda, Sepia officinalis, is more often represented by its calcareous dorsal plate than by the entire animal, for large numbers of these "cuttle-bones" are sometimes strewed along the shore for miles. The Squid, Loligo vulgaris, is often met with, sometimes of considerable size. The horny "pen" resembles a short leaf-shaped Roman sword, and Browne's term, "Gladiolus," is quite as appropriate as that of "Calamus." His Polypus is probably Octopus vulgaris, but it is rarely met with on the Norfolk coast.

The Loligo sleue or calamar found often upon the shoare from head to tayle [such crossed out] sometimes aboue an ell long, remarkable for its parretlike bill, the gladiolus or calamus along the back & the notable crystallyne of the eye wch equalleth if not exceedeth the lustre of orientall pearle.

A polypus another kind of the mollia[N] sometimes wee haue met with.

[N] By mollia is meant all soft-bodied shell-less animals.

Lobsters in great number about sheringham and cromer from whence all the country is supplyed.

Astacus marinus pediculi [marini written above] facie[81] found also in that place. with the aduantage of ye long foreclawes about 4 inches long.

[81] Probably Nephrops norvegicus, the Norway Lobster, called at Lowestoft a Crayfish or Prawn. They are sometimes brought in in large numbers by the steam trawlers, but the precise locality in which they are captured I am unable to say; the fishermen say the "North Sea," which is rather a vague address, but

others say between the Texel and Heligoland.

Crabs large & well tasted found also in the same coast.

Another kind of crab[82] taken for cancer fluuiatilis litle slender & of a very quick motion found in the Riuer running through yarmouth. [added subsequently] & in bliburgh riuer.

[82] Carsinus maenas, the Shore-crab, a very common species on the Norfolk coast is here intended.

[Fol. 33.] Oysters exceeding large about Burnham and [Huns crossed out] Hunstanton like those of poole St Mallowes or ciuita [vech crossed out] vechia whereof [some crossed out] many are eaten rawe the shells being broakin with [cle crossed out] cleuers the greater part pickled & sent weekly to London & other parts.

Mituli or muscles in great quantitie as also chams or cochles about stiskay [sic] & ye northwest coast.

Pectines pectunculi varij or scallops of the lesser sort.

Turbines or smaller wilks, leues, striati. as also Trochi, Trochili, or scaloppes finely variegated & pearly. [as also crossed out.] Lewise [sic] purpurae minores, nerites, cochleae, Tellinae.

Lepades, patellae Limpets, of an vniualue shell wherein an animal like a snayle cleauing fast unto the rocks.

Solenes cappe lunge venetorum co[=m]only a razor fish the shell thereof dentalia

[The MS. breaks off here, and the next paragraph appears to be an interpolation.]

Dentalia by some called pinpaches because pinmeat thereof is taken out with a pinne or needle.[83]

[83] Mussels and Cockles are very abundant all along the shallow shores of North-west Norfolk, as well as Clams, Mya arenaria. "Scallops of the lesser sort" are probably Pecten opercularius and P. varius. The Whelk, Buccinum undatum, is also very numerous, and forms the staple of a considerable industry at Sheringham; the lesser, or Dog-Whelk, Nassa reticulata, as well as Purpura lapillus and several sorts of Trochus, are commonly met with. The genus Nerita was a very comprehensive one in Browne's time, and included many species of Littorina, of which the well-known Periwinkle, L. littorea, is the most numerous here. No true Nerita is now recognised as British, although in the warmer seas the genus is a very numerous one. The most common Tellina here is T. tenuis, Lepades patellae are of course the common Limpet (Patella vulgata), and of the Solen, or Razor Shell, which Gwyn Jeffreys says in the time of Aldrovandus was called by the Venetians "cappa longa," we have two species found on the sandy portions of the coast. Here some confusion exists in the MS., after the words, "the shell thereof dentalia," the note ends abruptly, and is followed by an interpolation which seems quite irrelevant, as Dentalia have surely never been called "Pin-patches" (the vernacular name for Littorina littorea), nor is it probable that, like that common univalve, they were ever taken out of their shells with a pin or needle. Dentalia are mentioned on two other occasions as of doubtful occurrence and Dentalium entalis has slight claim to be a native of Norfolk; the only recorded specimen I know of was picked up in 1890 by Mr. Mayfield, from the drift on the beach between Wells and Holkham.

Cancellus Turbinum et neritis[84] Barnard the Hermite of Rondeletius a kind of crab or astacus liuing in a forsaken wilk or nerites.

[84] Hermit Crabs are here referred to, the larger, Pagurus bernhardus, found very frequently inhabiting the shells of the Whelk, and a smaller species which takes up its abode in those of a Trochus.

echinus echinometrites[85] sea hedghogge whose neat shells are co[=m]on on the shoare the fish aliue often taken [with crossed out] by the dragges among the oysters.

[85] Dead Echini are very common on the sea-shore, and many living ones are dredged by the shrimpers. Echinus sphaera is the most common on the Norfolk coast; E. miliaris, a small species, is also very abundant about Cromer.

[This and the next paragraph on fol. 33 verso.]

Balani[86] a smaller sort of vniualue growing co[=m]only in clusters. the smaller kinds thereof to bee found oftimes upon oysters wilks & lobsters.

[86] The species of Cirripeds referred to are probably the common Acorn Barnacle (Balanus porcatus) and the Goose Barnacle (Lepas anatifera), the latter occasionally found on ships' bottoms and drift-wood, probably carried by favourable currents from warmer seas than our own.

Concha anatifera or Ansifera or Barnicleshell whereof about 4 yeares past were found upon the shoare no small number by yarmouth hanging by slender strings of a kind of Alga vnto seuerall splinters or [clefts crossed out] cleauings of firre boards vnto wch they were seuerally fastned & hanged like ropes of onyons: their shell flat & of a peculiar forme differing from other shelles, this being of four diuisions. containing a small imperfect animal at the lower part diuided into many shootes or streames wch prepossed [imag crossed out] spectators fancy to bee the rudiment of the tayle of some goose or duck to bee [expute crossed out] produced from it. some whereof in ye shell & some taken out & spred upon paper wee shall [still?] keepe by us.

[Fol. 34.] Stellae marinae[87] or sea starres in great plentie especially about yarmouth. whether they bee bred out of the [vrticas crossed out] vrticae squalders or sea gellies as many report wee cannot confirme butt the squalderes in the middle seeme to haue some lines or first draughts not unlike.

our starres exceed not 5 poynts though I haue heard that some with more haue been found about Hunstanton and Burnham. where are also found stellae marinae testacae or handsome crusted & brittle sea [stars crossed out] starres much lesse.

[87] The Five-finger (Asterias rubens, L.) is a very numerous species on our coast and very destructive. Brittle Stars (Ophiocoma sp?) are as Browne states most frequent about Hunstanton, Burnham, and Cromer. Solaster papposa is also found in the same localities.

The pediculus[88] and culex marin us the sea lowse & flie are [are crossed out] also no strangeres.

[88] The Pediculus, or Sea Louse, is probably Talitrus locusta, the Sand-hopper; what may be intended by Culex marinus it is difficult to say. A species of gnat is at times very numerous on the wet sand just above the water-line. See also Notes 110 and 115, on a kindred subject.

Physsalus Rondeletij[89] or eruca marina physsaloides according to the icon of Rondeletius of very orient green & purple bristles.

[89] The Sea Mouse, Aphrodite aculeata. This is referred to again in the Letters to Merrett.

Urtica marina[90] of diuers kinds some whereof called squalderes. of a burning and stinging qualitie if rubbed in the hand. the water thereof may afford a good cosmetick.

[90] Mr. E. T. Browne, of the Zoological Laboratory of University College, London, has kindly furnished me with the following notes on this subject: "Jonston (1657) gives figures of Anemones and large Medusae under the name of Urtica. On Tab. xviii. he figures Anemones and other beasts, but not medusae. The medusae are on the next Tab. (xix.). Urtica marina includes both Anemones and certain Scyphomedusae (not Pulmo). Under 'some ... called

Squalders of a burning and stinging quality,' I think Browne must refer to our common stinging Scyphomedusae belonging to the genus Chrysaora or Cyanaea, of which there are three species.

"The vague description of what he calls 'sea buttons' [see below, also second letter to Merrett] would suit either a Medusa or a Ctenophore. The additional note, 'two small holes in the ends,' rather upsets matters, but I think he must refer to some sort of jelly-fish, probably damaged, which is usually the case when cast up on the shore. If the buttons worn in those days were like filbert-nuts or eggs, I am inclined to think that the reference must be to a Ctenophore, genus Pleurobrachia, but if flat, then to one of the Hydromedusae. It would be safe to say, 'probably a kind of jelly-fish,' which is about as vague as the reference." See also Dr. Reuben Robinson's description of "Squalders" in a letter to Browne (Wilkin i., pp. 422-424). It seems probable that the gelatinous masses referred to in the early part of this letter, which Dr. Robinson says were ascribed by Dr. Charleton to "the nocturnall pollution of some plethorick or wanton starr: or rather excrement blowne from the nosthrills of a rheumatick planett," were the remains of the undeveloped spawn of frogs, the bodies of which had been eaten by rats, crows, or herons, and which had become swollen by exposure to moisture.

[The next paragraph on folio 33 verso is evidently added subsequently.]

Another elegant sort that is often found cast up by shoare in great numbers about ye bignesse of a button cleere & welted & may bee called fibula marina crystallina.

hirudines marini or sea Leaches.[91]

[91] It is difficult to determine the species of marine Annelids referred to by Browne; the Sea Leech is probably Pontobdella laevis. The "large wormes" digged for bait, mentioned more than once, are Lug-worms, Arenicola piscatorum; the Vermes in tubulis testacei may be tube-worms of the genus Terrebella, or a species of Serpula. Tethya or "Sea dugge" (not "Sea dogs," as

Wilkin has it) might very well apply to Ascidia or one of the allied genera. Simple Ascidians, generally known as Sea-squirts, are common littoral forms; the animals figured by Rondeletius under the heading "De Tethyis" (p. 127) are simple Ascidians. The vesicaria marina, or "fanago," might well refer to the egg capsules of the common Whelk (Buccinum undatum), which are very commonly found in masses on the shore. In his sixth letter to Merrett, Browne mentions two kinds of "fanago," the first which I take to be the egg capsules of the Whelk, resembling the "husk of peas;" the smaller that of "barley when the flower [awn?] is mouldered away," may possibly be the egg capsules of Purpura lapillus, or of some species of Natica, which bear a fanciful resemblance to grains of barley. See also Merrett's second letter in Appendix A., in which he describes the Vesicaria found on oyster-shells as resembling flowers of Hyacinthus botryoides, which is not a bad description of the form of the egg capsules of P. lapillus.

vermes marini very large wormes digged a yarde deepe out of the sands at the ebbe for bayt. tis known where they are to bee found by a litle flat ouer them on ye surface of ye sand. as also vermes in tubulis testacei. Also Tethya or sea dugges some whereof resemble fritters [and crossed out] the vesicaria marina also & [see Note 91] fanago sometimes very large conceaued to proceed from some testaceous animals. & particularly [Fol. 35] from the purpura butt [in crossed out] ours more probably from other testaceous wee hauing not met with any large purpura upon this coast.

[A blank space.]

Many riuer fishes also and animals. Salmon[92] no co[=m]on fish in our riuers though many are taken in the owse. in the Bure or north riuer, in ye waueney or south riuer, in ye [yare or crossed out] norwich riuer butt seldome and in the winter butt 4 yeares ago 15 were taken at Trowes mill [ab crossed out] in Xtmas. whose mouths were stuck with small wormes or horsleaches no bigger than fine threads some of these I kept in water 3 moneths if a few drops of blood were putt to the water they would in a litle time looke red. they sensibly grewe bigger then I first found them and were killed by an hard froast freezing the water. most of our Salmons haue a recurued peece of flesh in ye end of the lower iawe wch when they shutt there mouths deepely enters the upper. as Scaliger hath noted in some.

[92] The Salmon (Salmo salar) is at the present day very rarely found in our rivers, and those met with are, as a rule, male Kelts which have strayed into unsuspected situations after floods; a singular exception occurred on the 20th May, 1897, when one weighing 6 lbs. was taken on a fly in the river above Stoke Holy Cross Mill; this fish is preserved in the Norwich Museum. Another curious capture of which I heard (but did not see the fish) occurred on the 1st August, 1898, when a salmon, also of 6 lbs. weight, jumped into a small boat towed behind a yacht which was sailing across Breydon Water. That the salmon was at one time a recognised visitor to our rivers is evident from the following extract from the Norwich Court of Mayoralty Book under date 2 Novr. 1667: "It is ordered that the bell man give notice that if any person shall take any Salmons from the Nativity of our Lady unto St. Martin's day, or destroy any young Salmons by netts or other ingens from the midst of April until the Nativity of St. John Baptist shall be punished according to the law." The Salmon is the host of several parasites both internal and external. Fresh run Salmon are generally infested with a "Sea-louse," which quickly perishes in freshwater; not so, however, with the troublesome worm-like creature, the subject of Browne's experiments; it is known as Lernaea salmonis, and is only found on the gill-covers of spent Kelts; it is not got rid of till the fish returns to the salt water. Browne may be excused being rather sceptical as to the identity of the clean run Salmon and the spent Kelt, for no greater contrast can be imagined than that which exists between the two--the male in the "redding" season develops the unsightly hooked mandible, which so puzzled the worthy doctor, and both in colour and form is as hideous an object as can be imagined. Becard Gallorum (not Beccard gallorus), i.e., the fish called "Becard" by the French (see second letter to Merrett), refers to the use of a name still applied in France to a large Cock Salmon, and "Anchorago" is the name under which the fish was described by Scaliger, whose book I have not seen. Dr. Guenther tells me that Artedi, "Ichthyologia," Pt. v., p. 23, quotes this name as a synonym of the Salmon.

The Riuers lakes & broads[93] abound in [the Lucius or added above] pikes of very large size where also is found the Brama or [breme crossed out] Breme large & well tasted the Tinca or Tench the Rubecula Roach as also Rowds and Dare or Dace perca or pearch great & small. whereof such [as] are are in Braden on this side yarmouth in the mixed water [are gen crossed out] make a dish very dayntie & I think scarce to bee bettered in England. butt the Blea[k] [Fol. 36] the chubbe the barbell [I haue not obserued in these riuers crossed out] to bee found in diues other Riuers in England I haue not obserued in these. As also fewer mennowes then in many other riuers.

[93] The freshwater fishes named in the next three paragraphs are so well known as to require few remarks. The Bream in our rivers and broads are very numerous and reach a large size, but of their esculent qualities I have had no personal experience; not so, however, with the Perch, which quite deserve Browne's high encomium. It is well known here that this fish shows no aversion of a certain admixture of salt and fresh water, and Mr. Lubbock ("Fauna of Norfolk") says, "the point in Norfolk rivers where the largest are taken with most certainty is where water begins to turn brackish from the influence of the ocean;" in autumn the very finest are taken by angling with a shrimp, a favourite bait in the lower parts of the Yare and Waveney. In such localities a small shrimp (Hippolyte varians, Leach) abounds, and it is to this favourite food that Mr. Lubbock attributes the excellence of these Perch. Roud is the local name of the Rudd (Leuciscus erythropthalmus). The River Nar is still perhaps the best Trout stream in the county, and the Crawfish is found in most of the rivers but not abundantly.

The Trutta or trout the Gammarus or crawfish [no crossed out] butt scarce in our riuers butt frequently taken in the Bure or north riuer & in the seuerall branches therof. & very remarkable large crawfishes to bee found in the riuer wch runnes by castleaker & nerford.

The Aspredo perca minor[94] and probably the cernua of Cardan co[=m]only called a Ruffe in great plentie in norwich Riuers & euen in ye streame of the citty. which though camden appropriates vnto this citty yet they are also found in the riuers of oxforde [&] Cambridge.

[94] Merrett calls the Ruff Cernua fluviatilis, and mentions its abundance in the River Yare at Norwich, which he (no doubt inadvertently) assigns to the County of "Essex"; from this locality Caius obtained the specimen, a drawing of which he sent to Gesner under the name of Aspredo. Camden assigns this fish also to Norwich, and Spencer, in his "Marriage of the Thames and Medway," writes of the Ruff:--

"Next cometh Yar, soft washing Norwich walls, And with him bringeth to their festival Fish whose like none else can show, The which men Ruffins call."

This county seems to have been assigned an exclusive proprietorship in the Ruff, to which, as Browne rightly points out, it had no just claim.

Lampetra Lampries great & small[95] found plentifully in norwich riuer & euen in the Citty about may [some crossed out] whereof some are very large & well cooked are counted a dayntie bitt collard up butt especially in pyes.

[95] Both the Sea Lamprey (Petromyzon marinus) and the Lampern (P. fluviatilis) are found in the Norfolk rivers.

Mustela fluuiatilis or eele poult[96] to bee had in norwich riuer & [in thalso crossed out] between it & yarmouth as also in the riuers of marshland resembling an eele & a cod. a very good dish & the Liuer thereof well answers the commendations of the Ancients.

[96] The Burbot, or Eel Pout (Lola vulgaris), called by Merrett a Coney-fish, from its habit of concealing itself in holes in the river banks. It is not sufficiently numerous now to form an article of diet, and I imagine there are few living who could bear testimony as to the esculent qualities of its "Liuer."

[Fol 37.] Godgions or funduli fluuiatiles, many whereof may bee taken within the [citty crossed out] Riuer in the citty:

Capitones fluuiatilis or millers thumbs, pungitius fluuiatilis or stanticles. Aphia cobites fluuiatilis or Loches. in norwich riuers in the runnes about Heueningham heath in the north riuer & streames thereof.

Of eeles[97] the co[=m]on eele & the glot wch hath somewhat a different shape in the bignesse of the head & is affirmed to have yong ones often found within it. & wee haue found a vterus in the same somewhat answering the icon thereof in Senesinus.

[97] The coarse variety of the Eel, known as the "Glout," or Broad-nosed Eel, is believed to be the barren female; Browne's informants were doubtless misled by the presence of certain thread-worms (Nematoxys) in the abdomen of the eels, which they mistook for young ones.

Carpiones carpes plentifull in ponds & sometimes large ones in broads [smear] 2 the largest I euer beheld were [found crossed out] taken [added above] in Norwich Riuer.

[A whole line is smeared out, and a break occurs in the MS. after the observation on the Carp; it then proceeds to notice some other inhabitants of the county which perhaps Browne had difficulty in classifying.]

Though the woods and dryelands about [abound?] with adders and vipers[98] yet are there few snakes about our riuers or meadowes more to bee found in Marsh land butt ponds & plashes abound in Lizards or swifts.

[98] Both Vipers (or Adders) and Snakes, the latter in particular, are, I imagine, much less abundant than formerly, but the few species of Lizards and Newts (Swifts) are still probably in undiminished numbers; the Mole Cricket (Gryllotalpa vulgaris) is rare with us; Horse-leeches (Aulostoma gulo) are frequent, and also "Periwinkles," which I take to be various species of

freshwater Molluscs, possibly of Limnaea. The Hard-worm (or Hair-worm), Gordius aquaticus, which refused to be generated from "horsehayres," is still an object of wonder to the unlearned, and the Great Black Water-Beetle (Hydrophilus piceus) is found; but forficula and corculum were a puzzle, as it is evident from their association they must be aquatic forms (and the Earwig certainly does not take to the water voluntarily), till my friend, Mr. C. G. Barrett, referred me to the following passage in Swammerdam's "Book of Nature," p. 93: "This is most certain that the Forficula aquatica of Jonston is the true nymph of the Mordella, or Dragon-fly,"[O] Dr. Charleton in his "Onomasticon," p. 57, has "Corculus, the Water-beetle, resembling an heart;" not very definite, but probably the Whirligig Beetle, Gyrinus natator, is intended; it is also an appellation given by some authors to "a small species of cordiformis, or heart-shell, of a rose colour," doubtless a Cyclas or a Pisidium. Squilla is the Freshwater Shrimp (Gammarus pulex), and Notonecta glauca, the Waterboatman "which swimmeth on its back," is well known.

Otters are still numerous in the broads and reed-margined rivers, and so long as these natural fastnesses endure in their present condition they are likely to continue so.

[O] On reference to Jonston (Historiae Naturalis de Insectis Lib. iv., "De Insectis aquaticis" i., p. 189, Tab. xxvii.), I find that under the name of "Forficulae aquat[icae]. M [oufet]," he has two figures, the first of which is possibly a Dytiscus larva, the second that of some form of Dragon-fly, which however is imperfect.

The Gryllotalpa or fencricket co[=m]on in fenny places butt wee haue met with them also in dry places dung-hills & church yards of this citty.

Beside horseleaches & periwinkles in plashes & standing waters we haue met with vermes setacei or hardwormes butt could neuer conuert horsehayres into them by laying them in water as also the [Fol. 38] the (bis) great Hydrocantharus or black shining water Beetle the forficula, sqilla, corculum and notonecton that swimmeth on its back.

Camden [smear] reports that in former time there haue been [otters crossed out] Beuers in the Riuer of Cardigan in wales. this wee are to sure of that the Riuers great Broads & carres afford great store of otters with us, a [des crossed out] great destroyer of fish as feeding butt from ye vent downewards. [a prey crossed out] not free from being a prey it self for their yong ones haue been found in Buzzards nests. they are accounted no bad dish by many are to bee made very tame and in some howses haue [semed crossed out] serued for turnespitts.

[Blank space.]

NOTE.--Although Browne's account of the Fishes is doubtless derived from his personal observation, I have found it very difficult in some families, such as the Cods, Rays, Gurnards, Flat-fishes, and Gobies to identify them with the species as at present known; in fact, they were at that time very imperfectly differentiated, and the figures in the old authors are generally so inexact as not to be recognisable. Ray, in 1674 ("English Words not generally known," p. 101), thus writes of the sea fishes, "several of them, we judge, not yet described by any Author extant in print: indeed the writers of Natural History of Animals living far from the Ocean, and so having never had opportunity of seeing these kind of fishes ... write very confusedly and obscurely concerning them," a remark which I have found abundantly verified.

LETTERS TO MERRETT.

[MS. SLOANE. 1833. FOL. 14.]

No. 1.

"My father to Dr. Meret July 13, 1668."

Most honourd Sir,

[Fol 14.] I take ye boldnesse to salute you as a person of singular worth & learning and whom I very much respect & honour. I presented my service to you by my sonne some months past, and had thought before this time to have done it by him again, but the time of his returne to London being yet uncertaine, I would not deferre these at present unto you. I should be very glad to serve you by any observations of mine against yr. second edition of your Pinax[99] which I cannot sufficiently commende. I have observed and taken notice of many animals in these parts whereof 3 years agoe a learned gentleman of this country desired me to give him some account, which while I was doing ye gentleman my good friend died. I shall only at this time present and name some few unto you which I found not in your catalogue. A Trachurus [see Note 61] which yearly cometh before or in ye head of ye herrings called therefore an horse. Stella marina testacea [see Note 87] which I have often found upon the sea-shoare, an Astacus marinus pediculi marini facie [see Note 81] which is sometimes taken with the lobsters at Cromer in Norfolck. a pungitius marinus [see Note 75] wereof I have known many taken among weeds by fishers who drag by ye Sea-shoare on this coast. A Scarabaeus capricornus odoratus[100] which I take to be mentioned by Moufetus fol. 150. I have taken some abroad one in my Seller which I now send he saith nucem moschatam et cinamomum vere Spirat to me it smelt like roses santalum & Ambegris. I have thrice met with Mergus maximus Farensis Clusij, [see Note 11] and have a draught thereof. they were taken about the time of herring fishing at yarmouth one was taken upon the shoare not able to fly away about ten yeares agoe I sent one to Dr. Scarborough. Twice I have met with a Skua Hoyeri [see Note 10] the draught whereof I also have. one was shot in a marsh which I gave unto a gentleman which [sic] I can sende you another was killd feeding upon a dead horse neere a marsh ground. Perusing your catalogue of Plants. upon Acorus verus,[101] I find these wordes found by Dr. Browne neere Lin. wherein probably there may be some mistake, for I cannot affirme nor I doubt any other yt. is found thereabout. Some 25 yeares ago I gave an account of this plant unto [this crossed out] Mr. Goodyeere:[102] & more lately to Dr. How[103] unto whome I sent some notes and a box full of the fresh Juli. This elegant plant groweth very plentifully and beareth its Julus yearly by the bankes of Norwich river [fol. 13 verso] chiefly about Claxton and Surlingham. & also between norwich & Hellsden bridge so that I have known Heigham Church in the suburbes of Norwich strowed all over with it, it hath been transplanted and set on the sides of Marish pondes in severall places of the country where it thrives and beareth ye Julus yearly.

[99] It is evident that Merrett was collecting a considerable amount of materials for an enlarged edition of his Pinax Rerum Naturalium Britannicarum, on behalf of which Browne seems, by this introductory letter, to have tendered his assistance, but the contemplated edition, probably for reasons which I have mentioned elsewhere, never appeared; happily, these rough drafts have been preserved, although it seems not unlikely that the letters themselves, should they ever be found, would differ from them in some respects.

[100] Scarabaeus capricornus odoratus. The Musk Beetle, Aromia moschata, L.

[101] Acorus calamus, the Sweet Flag, is still found in plenty in various localities in the county, but it does not appear to develop its curious "julus" every year. It was very abundant at Heigham, a suburb of Norwich, on the site now occupied by the goods yard of the Midland and Great Northern Railway, and it was probably from this spot that the supply was obtained for the purpose of littering the floor of the old parish church. Mr. Vaux, in his "Church Folk-Lore," p. 264, says that up to the passing of the Municipal Reform Bill the Town Clerk of Norwich was accustomed to pay the sub-sacrist of the cathedral an amount of one guinea for strewing the floor with rushes on the Mayor's Day. The custom is said to have been adopted "as well for coolness as for pleasant smell." The pleasant cinnamon-like scent of the rush, on being trodden on, is said to have perfumed the whole building. The root was also used as a remedy in cases of ague, and formed the base of tooth and hair powders.

[102] Towards the end of the Introductory Letter to Johnson's (1636) Edition

of Gerard's "Herball," he acknowledges the assistance he received from Mr. John Goodyer, of Maple-Durham, in Hampshire. Sir J. E. Smith ("Eng. Flora," iv., p. 34) speaks of him as "one of the most deserving of our early English Botanists." Robert Brown named a genus of plants (Goodyera) after Goodyer.

[103] William How, 1620-1656, was the author of "Phytologia Britannica," Lond., 1650, "the earliest work on botany restricted to the plants of this island" ("Dic. of Nat. Biog."). He practised medicine in London.

Sesamoides Salamanticum Magnum.[104] Why you omit Sesamoides Salamanticum parvum this groweth not far from Thetford and Brandon and plentifull in neighbour places where I found it and have it in my hortus hyemalis answering ye description in Gerard.

[104] Sesamoides is stated in Ree's Encyclopaedia and in Eng. Fl. to be a synonym of Reseda, therefore Sesamoides magnum would appear to be R. luteola and S. parvum, R. lutea.

Urtica Romana[105] which groweth with button seede bags is not in yr. catalogue I have founde it to grow wild at [Golston crossed out] Golston by Yarmouth, & transplanted it to other places.[P]

[105] Urtica Romana, which is again referred to as U. mas near the end of the third letter and as being found at Gorleston, is the Roman Nettle, U. pilulifera. In 1834 the Pagets ("Nat. Hist. of Great Yarmouth") reported it as still found under old walls at Gorleston, "but rarer than formerly," and it is only in recent years that it has been exterminated, owing to building operations in that locality.

[P] This letter, evidently a copy as shown by the heading "My father to Dr. Meret," is in the writing of Dr. Edwd. Browne.

[MS. SLOANE 1830. FOL. 39-40.]

No. II.

Fol. 39.]

"My second letter to Dr Meret Aug xiiii 1668."

Honord Sr I receiued your courteous letter & am sorry some diuersions have so long delayed this my second vnto you. You are very exact in the account of the fungi. I have met with two,[106] which I have not found in any Author, of which I have sent you a rude draught inclosed. The first an elegant fungus Ligneus found in an hollow sallowe I haue one of them by mee butt without a very good opportunitie dare not send it fearing it should bee broken vnto some it seemed to resemble some noble or princely ornament of the head & so might bee called fungus Regius vnto others a turret, top of a cupola or Lanterne of a building & so might bee named fungus pterygoides, pinnacularis or Lanterniformis you may name it as you please. The second fungus Ligneus teres Antliarum or fungus ligularis longissimus consisting [of crossed out] or made of many wooddy strings about the bignesse of round poynts or Laces some about half a yard long shooting in a bushie forme from the trees wch serue vnderground for pumpes. I have obserued diuers especially in norwich where wells are sunck deep for pumpes.

[106] Dr. Plowright informs me that "it is impossible to say with certainty what the first named Fungus is; the description suggests some form of Polyporus perhaps, P. varius, which is a ligneous species and occurs frequently on willows in Norfolk. The second is the abortive form of Polyporus squamosus, which is well figured by many of the older botanists, for instance under the name of Boletus rangiferinus, by Bolton, t. 138, and Boletus squamosus, var. rangiferinus, by Hooker, 'Flora Londinensis,' new series. In many cases no pileus at all is formed and it used then to be referred to Clavaria." The Phalloides is Phallus impudicus, L., a very common species in this county and even occurring in some of the city gardens where its exceedingly offensive odour renders it very undesirable. Fungus rotundus is the well-known Lycoperdon giganteum, Fr., which sometimes reaches a very

large size.

The fungus phalloides found not farre from norwich large & very fetid answering the description of Hadrianus junius I have a part of one dryed by mee.

Fungus rotundus maior I haue found about x inches in Diameter & half [sic, have?] half a one dryed by mee.

Another small paper containes the rude draughts of fibulae marinae pellucidae, [see Note 90] or sea buttons a kind of squalder & referring to vrtica marina which I haue obserued in great numbers by yarmouth after a flood & easterly winds. They resemble pure crystall buttons chamfered or welted on the sides with 2 small holes at the ends. They cannot bee sent for the included water or thinne gelly soon runneth from them.

Vrtica marina minor jonstoni [see Note 90] I haue often found on this coast. [Continued on fol. 39 verso.]

Physsalus [see Note 89] I haue often found also I haue one dryed but it hath lost its shape & colour.

Galei & caniculae [see Note 56] are often found I haue a fish hanged up in my yard of 2 yards long taken among the Herrings at yarmouth which is the Canis carcharias alius Johnstoni. Tab. vi fig. 6.

Lupus marinus you mention upon an handsome experiment butt I find it not in the catalogue. This Lupus marinus or Lycostomus is often taken by our seamen wch fish for cods I haue had diuers brought mee. they hang up in many howses in Yarmouth.

Trutta marina is taken with us--a better dish than the Riuer trowt butt of the same bignesse.

Loligo sepia a cuttle page 191 of your Pinax [see Note 80] I conceiue worthy Sr it were best to putt them in 2 distinct lines as distinct species of the Molles. The loligo, calamare or sleue I haue often found cast up on the seashoare & some haue been brought mee by fishermen of aboue [20 crossed out] twentie pound wayet.

Among the fishes of our Norwich riuer wee scarce reckon salmons [see Note 92] yet some are yearly taken. butt all taken in the Riuer or coast haue the end of the lower jaw very much hooked which enters a great way into the upper jaw like a socket. you may find the same though not in figure if you please to read Johnstonus fol 101 I am not satisfied with the conceit of some authors there that is [it?] is a difference of male and female for all ours are thus formed. The fish is thicker than [oth crossed out] ordinarie salmons and very much & more largely spotted whether not rather Beccard gallorum or Anchorago Scaligeri I haue bothe draught & head of one dryed either of wch you may command.

Scyllarus or cancellus in turbine tis probable you have [see Note 84]. haue you cancellus in nerite a small testaceous found upon this coast.

[Fol. 40.] Haue you mullus ruber asper [see Note 63].

Haue [you] piscis octangularis Bivormii?[Q] [see Note 66, also pp. 65 and 87 infra].

[Q] Thus in the MS., but Browne seems to have intended to write Bicornis Vormii, and accidentally to have run the two words together [see p. 41 supra].

vermes marini larger than earthwormes [see Note 91] digged out of the sea sand about 2 foot deepe at an ebbe water for bayte they are discouered by a little hole or sinking of the sand at the top aboue them.

Haue you that handsome colourd [bird crossed out] jay [see Note 49] answering the description of Garrulus Argentoratensis & may be called the parret jay I haue one that was killed upon a tree about 5 yeares ago.

Haue you a may chitt a small dark gray bird [see Note 29] about the bignesse of a stint wch cometh about may & stayeth butt a moneth. a bird of exceeding fattnesse and accounted a daintie dish. they are plentifully taken in marshland and about wisbich.

Haue you a [caprimulgus or written above] dorhawke a bird as bigge as [a] pigeon [see Note 42] with a wide throat bill as little as a titmous & white fethers in the tayle & paned like an hawke.

Succinum raro occurrit[107] pag 291 of yours. [Should be p. 219] not so rarely on the coast of norfolk. tis usually found in small peeces [butt crossed out] sometimes in peeces of a pound wayght. I haue one by mee fat & fayre of x ounces wayght--jet more often found I haue an handsom peece of xii ounces in wayet.

[107] Amber, writes Mr. Clement Reid, in a paper contributed by him to the "Trans. Norf. and Nor. Nat. Soc." (iii., p. 601), "is found on the Norfolk coast, usually mixed with the seaweed thrown up by the Spring gales," but is very rarely found in place; as much as three or four pounds are annually gathered near Cromer. The quality, Mr. Rein says, is very good, but the dark transparent lumps are most generally found. In a subsequent paper (op. cit., iv., p. 248) he enumerates seven species of insects which have been found enclosed, and in a third communication mentions an eighth. Mr. A. S. Ford, as the result of an examination of a collection of East-coast Amber made at Yarmouth (op. cit., v., p. 92), adds one species of Hymenoptera, three of Coleoptera, two of Orthoptera, with some Araneida, and remains of vegetable substances which had not been identified.

The Jet found on the Norfolk coast differs considerably from the Whitby Jet, and Mr. Reid, "Geology of the Country Round Cromer" (p. 133), believes that in all probability it was originally derived from Lower Tertiary beds under the North Sea, a few miles from the present coast. Mr. Savin estimates the average annual find of Jet near Cromer at from ten to twenty pounds.

The doctor does not display his usual acumen when he rejects the "ancient" opinion as to the vegetable origin of Amber, see Pseudodoxia, book ii., chap. iv.; also letter from Earl of Yarmouth to T. B. (Wilkin Edit. i., p. 411).

No. III.

[FOL. 40 verso.]

"My third letter Sept xiii."

Sr I receaued your courteous Letter and with all respects I now agayne salute you.

The mola piscis is almost yearely taken on our coast [see Note 58] this [last crossed out] year one was taken of about 2 hundred pounds wayght diuers of them I haue opened & haue found many lyce sticking close vnto thier gills whereof I send you some.

In your pinax I find onocrotalus or pellican [see Note 25] whether you meane those at St. James or others brought ouer or such as haue been taken or killed heere I knowe not. I haue one hangd up in my howse wch was shott in a fenne ten miles of about 4 yeares ago and because it was so rare some conjectured it might bee one of those which belonged vnto the King & flewe away.

Ciconia raro hue aduolat. I haue seen two [see Note 14] one in a watery marsh 8 miles of, another shott whose case is yet to bee seen. [See Appendix D.]

Vitulus marinus. In tractibus borealibus et Scotia [see Note 53]. no raritie upon the coast of Norfolk at a lowe water I haue knowne them taken asleep vnder the cliffes. diuers haue been brought vnto mee. our seale is different from the Mediterranean seale. as hauing a rounder head a shorter and stronger body.

Rana piscatrix I haue often known taken on our coast & some very large [see Note 59].

Xiphias or gladius piscis or sword fish wee haue in our seas [see Note 55]. I haue the head of one which was taken not long ago entangled in the Herring netts the sword aboue 2 foot in length.

Among the whales you may very well putt in the spermacetus [see Note 51] or that remarkably peculiar whale which so aboundeth in spermaceti. about twelve years ago wee had one cast up on our shoare neer welles wch I discribed in a peculiar chapter in the last edition of [Fol. 41] my pseudodoxia epidemica. another was diuers yeares before cast up at Hunstanton. both whose heads are yet to bee seen.

Ophidion or at least ophidion nostras [see Note 69] co[=m]only called a sting fish hauing a small prickley finne running all along the back, & another a good way on the belly, with little black spotts at the bottom of the back finne if the fishermens hands bee touched or scrached with this venemous fish they grow paynfull and swell the figure hereof I send you in colours they are co[=m]on about cromer see Schoneveldeus de Ophidiis.

Piscis octogonius or octangularis answering the discription of Cataphractus Schoneveldei [see Note 66] only his is discribed with the finnes spread & when it was fresh taken & a large one howeuer this may bee nostras I send you one butt I haue seen much larger which fishermen haue brought mee.

Physsalus [see Note 89]. I send one which hath been long opened & shrunck & lost the colour when I tooke it upon the sea shoare it was full & plump answering the figure & discription of Rondeletius. there is also a like figure at the end of [Rondeletius crossed out] muffetus I haue kept them aliue butt obserued no motion [butt crossed out] except of contraction and dilation when it is fresh the prickles or brisles are of a brisk green & Amethest colours--some call it a sea mous.

Our mullet is white & imberbis [see Note 63] butt wee haue also a mullis barbatus ruber miniaceus or cinnaberinus somewhat rough & butt drye meat. there is of them maior & minor resembling the figures in Johnstonus tab xvii Rotbart.

Of the Acus marinus or needle fishes [see Note 64] I haue obserued 3 sorts. The Acus Aristotelis called heere an Addercock Acus maior or Garfish with a green verdigris backbone the other saurus Acui similis Acus sauroides or sauriformis as it may be called much answering to the discription of saurus Rondeletij in the hinder part much resembling a makerell opening one I found not the backbone green Johnstonus writes nearest to it in his Acus minor. I send you the head of one dryed butt the bill is broken I haue the whole draught in picture. this kind is more rare then the other wch are co[=m]on & is a rounder fish.

[Fol. 41 verso.] Vermes marini are large wormes [see Note 91] found 2 foot deep in the sea sands & are digged out at an ebbe for bayt.

The Avicula Maialis or may chitt [see Note 29] is a litle dark gray bird somewhat bigger then a stint which co[=m]eth in may or the later end of April & stayeth about a moneth. A marsh bird the legges & feet black without an heele the bill black about 3 quarters of an inch long they grow very fatt & are accounted a dayntie dish.

A Dorhawke a bird not full so bigge as a pigeon [see Note 42] somewhat of a woodcock colour & paned somewhat like an hawke with a bill not much bigger then that of a Titmouse [& very wide throat added above] known by the name of a dorhawke or prayer upon beetles, as though it were some kind of accipiter muscarius. in brief this accipiter cantharophagus or dorhawke [a word smeared out] is Avis Rostratula gutturosa, quasi coaxans, scarabaeis vescens, sub vesperam volans, ouum speciosissim[=u] [word smeared] excludens. I haue had many of them & am sorry I have not one to send you I spoake to a friend to shoote one butt I doubt they are gone ouer.

of the vpupa [see Note 35] diuers have been brought mee & some I haue obserued in these parts as I trauuyled about.

The Aquila Gesneri I sent [aliue added above] to Dr. Scarburg [see Note 3] who told mee it was kept in the colledge it was brought mee out of Ireland. I kept it 2 yeares in my howse I am sorry I haue only one fether of it to send you.

A shooing horn or Barker from the figure of the bill & barking note [see Note 38] a long made bird of white & blakish colour finne footed, a marsh bird & not rare some times of the yeare in marshland. it may upon vewe bee called Recuruirostra nostras or Auoseta much resembling the Auosettae [species crossed out] species in Johnstonus tab (54). I send you the head in picture

[A smeared out] stone curliews I haue kept in large cages [see Note 37] the[y] haue a prettie shrill note, not hard to bee got in some parts of norfolk.

[Fol. 42] Haue you Scorpius marinus Schoneueldei [see Note 68]

haue you putt in the musca Tulipar[=u] muscata[108]

[108] It seems impossible to identify this insect; Merodon narcissi has been suggested, but Mr. Verrall, whom I consulted says, "certainly not Merodon, which probably was not known in Britain until about 1870," and suggests the small fly Nemopoda. Mr. Bloomfield writes that the only fly of which he has seen any mention as having a musky or "excellent fragrant odour" is Sepsis cynipsea, which Kirby and Spence state on the authority of De Geer, "emits a fragrant odour of beaum" (balm); this species is very nearly allied to Nemopoda. Several Bees, for instance the Genus Prosopis, emit a strong scent of balm, and it is possible that Browne may have used the term "fly" in what is even now a popular sense, and that really some species of Bee may have called forth his remarks. It will be noticed that at p. 74 he speaks of it as a "small beelike flye."

That bird which I sayd much answered the discription of Garrulus Argentoratensis [see Note 49] I send you it was shott on a tree x miles of 4 yeares ago. it may well bee called the Parret Jay or Garrulus psittacoides speciosus. the colours are much faded. if you haue it before I should bee content to haue it agayne otherwise you may please to keep it.

Garrulus Bohemicus[109] probably you haue a prettie handsome bird with the fine cinnaberin tipps of the wings some wch I haue seen heere haue the tayle tipt with yellowe wch is not in the discription.

[109] Mr. Stevenson, whom very little relating to Norfolk Ornithology escaped, was well acquainted with Sir Thomas Browne's works, yet has in his "Birds of Norfolk" unaccountably overlooked this passage, and remarks that Browne does not appear to have noticed this species; he however not only refers to it as above, but evidently describes it from his personal observation. It is a very uncertain winter visitor to this county, but on rare occasions makes its appearance in considerable flocks. A remarkable instance of this occurred in the winter of 1866-7, when Mr. Stevenson, as the result of the examination of a very large series, contributed an exhaustive paper on the plumage of this handsome bird to the "Transactions of the Norf. and Nor. Nat. Soc.," iii., pp. 326-344.

I haue also sent you urtica mas [see Note 105] which I lately gathered at Golston by yarmouth where I found it to growe also 25 yeares ago. of the stella marina Testacea which I sent you [see Note 87] I do not find the figure in any booke.

I send you a few flies[110] which some unhealthful yeares about the first part of september I haue obserued so numerous upon plashes in the marshes & marish diches that in a small compasse it were no hard matter to gather a peck of them I brought some what my box would hold butt the greatest part are scatterd lost or giuen away for memorie sake I writ on my box muscae palustres Autumnales [See Appendix D.]

[110] Mr. Verrall assures me that even in the present day it is quite impossible to recognise the species of Diptera described by persons unacquainted with the particular group, and that Browne's remarks would apply to hundreds of species. It is possible that an Ephydra may be meant. This genus of small flies, says Mr. Verrall, abounds in such places as Browne describes, but it is likely that other species were with them.

worthy Sr I shall be euer redie to serue you who am Sr your humble Seruant

THO BROWNE.

Norwich, Sep 16. 1668.

No. IV.

"The fourth Letter to Dr. Merrett Decemb xxix." [1668]

[Fol. 42 verso.] Sr I am very joyfull that you haue recouered your health whereof I heartily wish the continuation for your own and the publick good. And I humbly thank you for the courteous present of your booke.[111] with much delight and satisfaction I had read the same not once in English I must needs acknowledge your co[=m]ent more acceptable to me then the text which I am sure is an hard obscure peice without it. though I haue not been a stranger unto the vitriarie Art both in England and abroad.

[111] This evidently refers to the gift of a copy of Merrett's Latin translation of Antonio Neri's L'Arte Vetraria (Firenze, 1612, 4to), published under the title of "The Art of Glass, translated into English with some observations on the Author," &c., in 1662, and a Latin edition in 1668.

I perceiue you haue proceeded farre in your Pinax. These few at present I am bold to propose & hint unto you intending God willing to salute you agayne.

A paragraph might probably be annexed unto Quercus. Though wee haue not all the exotick oakes, nor their excretions yet these and probably more supercrescences productions or excretions may bee obserued in England.

Viscum--polypodium--Juli pilulae-- Gemmae foraminatae [formicatae?] folior[=u]-- excrement[=u] fungosum verticibus scatens-- Excrementum Lanatum-- Capitula squamosa jacaeae aemula. Nodi--melleus Liquor--Tubera radicum vermibus scatentia--Muscus--Lichen-- Fungus--varae quercinae.[112]

[112] The Rev. E. N. Bloomfield has most kindly assisted me in attempting to identify the Parasitic products of the Oak mentioned above:

Viscum, is doubtless the Mistletoe.

Polypodium, the Common Polypody Fern.

Juli pilulae: "little balls on the flower catkins." The Currant Gall, Neurosterus baccarum, which is the spring form of N. lenticularis; Oliv.

Gemmae foraminatae [formicatae?] foliorum: "pimple-like buds on the leaves." Leaf-galls, such as the Silky Button, N. numismatis, Oliv., and the common Spangle, N. lenticularus, Oliv.

Excrementum fungosum verticibus scatens: "a spongy secretion bursting out from the ends of the shoots." The Oak Apple, Biorhiza terminalis, Fab.

Excrementum lanatum: the Woolly Gall, Andricus ramuli, L., a somewhat rare Gall, resembling a ball of cotton-wool.

Capitula squamosa jacaeae aemula: "little scaley (or imbricated) heads resembling the heads of Jacea" (Black Knapweed). The Artichoke Gall. Andricus fecundatrix; Htg.

Nodi: probably swellings of any sort, whether caused by insects or not.

Melleus liquor: Honey-dew, a secretion of Aphides.

Tubera radicum vermibus scatentia: "swollen tubers on the roots containing grubs;" without doubt the Root-Gall, Andricus radicis, Fab. Polythalamous Galls, often very large at the roots or on the trunk near the ground.

Mosses, Lichens, and Fungi, all "genuine products of the Oak," need no comment, but Mr. Bloomfield remarks, "How wonderfully observant Sir Thomas Browne must have been thus to distinguish the various galls, &c., and to point them out so distinctly."

Browne's contemporary, Dean Wren, seems sadly to have misunderstood the fructification of the Oak. In a note on Browne's remarks on the "Miseltoe" (Pseudodoxia, book ii., chap. vi.), he says, "Arboreous excrescences of the Oak are soe many as may raise the greatest wonder. Besides the gall, which is his proper fruite, hee shootes out oakerns, i.e., ut nunc vocamus (acornes), and oakes apples, and polypodye, and moss; five several sorts of excrescences." See also letter to his son, Dr. Edward Browne, in which Sir Thomas Browne says that "wee haue little or none of viscus quercinus, or miselto of the oake, in this country; butt I beleeve they have in the woods and parks of Oxfordshyre."--Wilkin, i, p. 279.

[Fol. 43.] Capillaris marina sparsa fucus capillaris marinus sparsus sive capillitius marinus or sea periwigge.[113] strings of this are often found on the sea shoare. but this is the full figure I haue seen 3 times as large.

[113] In Sir Thomas Browne's time the Hydrozoa were not distinguished from the Corallines, and both were regarded as vegetable growths. It is almost impossible to determine from his vague descriptions even to which section those mentioned belong, but although our exposed coast-line is not favourable to such growths, there are a few common species of Hydroid Zoophytes which abound here, and to these, fortunately, Browne's specimens appear to belong. What he calls the "Sea-perriwig" is doubtless Sertularia operculata, Lin.,

sometimes known as "Sea-hair," a very common and widely dispersed species.

I send you also [several words smeared out] a little elegant sea plant[114] which I pulled from a greater bush thereof which I haue resembling the back bone of a fish. Fucus marinus vertebratus pisciculi spinum referens Icthyorachius or what you thinck fitt.

[114] The little "Fucus," which he compares to the backbone of a fish, is probably Halecium halecinum, Lin., the "Herring-bone Coral" of Ellis, one of the most common Zoophytes on our coast. The "Abies," of which he suggests at p. 75 that this may be a "difference," is most likely Sertularia abietina, Lin., which this species resembles, but is less regularly pinnate; this may have led him to suppose that the "sprouts, wings, or leaves" may have fallen off. The Fucus marinus is most likely Fucus serratus.

And though perhaps it bee not worth the taking notice of formicae arenariae marinae or at least muscus formicarius marinus[115] yet I obserue great numbers by the seashoare and at yarmouth an open sandy coast, in a sunny day many large and winged ones may bee obserued upon & rising out of the [shoare crossed out] wet sands when the tide falls away.

[115] Swarms of Ants and Flies are no uncommon sight along the seashore at certain seasons of the year, and under the conditions which Browne describes. The Pagets ("Nat. Hist. of Great Yarmouth") mention that the fly, Actora aestuum, is common on the beach at high-water mark; but Mr. Verrall writes me that there are many others likely to be thus met with, such as Orygma luctuosa and Limosina zosterae, widely divergent species. In his "Journal of a Tour" into Derbyshire, Dr. Edward Browne, in crossing the sands of the Wash, mentions his satisfaction at the absence of the swarms of flies "with which all the fenne countrys are extremely pestered." See also Note 110 supra.

Notonecton an insect that swimmeth on its back [see Note 98] & mentioned by Muffettus may be obserued with us.

I send you a white Reed chock[116] by name some kind of Junco or litle sort thereof I haue had another very white when fresh.

[116] It is impossible to form an idea as to what is here intended. I know of no Juncus which would answer the description. Professor Newton reminds me that "Junco" was a common name for "a bird that inhabited reeds," and was loosely applied, some old authors taking it to be the Reed Thrush (i.e., the Great Reed-Warbler of these days), and others, the Reed-Sparrow or Bunting. But bearing in mind Browne's practice of referring to Jonston, it seems possible that the latter's Junco may be here intended, and that, as the figure (pl. 53) shows, is a small Sandpiper, almost certainly the Dunlin. It is lettered "Junco Bellonii," but this he must have taken second-hand from Aldrovandus, since Belon never used the word "Junco" in this connexion, but called it "Schoeniclus" or "Alouette-de-mer"--terms rendered Junco by Aldrovandus (iii. p. 487). Charleton took the same view in his "Onomasticon" (p. 108), published in 1668 (the year assigned as that of this letter), stating that it was so-called because "in juncis libenter degat," and identifying it with the Alouette-de-mer of the French, and the English "Stint, or Sparr, or Perr." Gilbert White appears to have thus applied the term (cf. "Life" by Rashleigh Holt-White, i. pp. 186, 194, 250). In one place he says, "No. five is Ray's Junco and the Turdus arundinaceus of Linn." That "Junco" is the name of a bird is absolutely certain, but the context, "very white when fresh," does not seem to admit of explanation.

Also the draught of a sea fowle called a sherewater [see Note 17] billed like a cormorant, feirce & snapping like it upon any touch. I kept 2 of them aliue 5 weekes cramming them with fish refusing of themselues to feed on anything & wearied with cramming them they liued 17 dayes without food. They often fly about fishing [ves crossed out] shipps when they cleans their fish & throwe away the offell. so that it may bee referred to the Lari as Larus niger gutture albido rostro adunco.

Gossander videtur esse puphini species [Pinax, p. 184]. worthy Sr that wch we call a gossander [see Note 19] & is no rare fowle among us is a large well coloured & marked diuing fowle most answering the [mer crossed out] Merganser. it may bee like the puffin in fattnesse and [Ranknesse crossed out] Ranknesse butt no fowle is I think like the puffin differenced from all others by a peculiar kind of bill

[Fol 43 verso.] Barganders [see Note 18] not so rare as Turn [Turner] makes them co[=m]on in Norfolk so abounding in vast & spatious warrens.

If you haue not yet putt in Larus minor or a sterne [see Note 13] it would not bee omitted, co[=m]on about broad waters and plashes not farre from the sea.

Haue you a Yarwhelp, Barker, or Latrator [see Note 39] a marsh bird about the bignesse of a Godwitt

Haue you Dentalia [see Note 83] which are small vniualue testacea whereof sometimes wee find some on the seashoare

Haue you putt in nerites another little Testaceum which wee haue [see Note 83].

Haue you an Apiaster a small bird calld a Beebird.[117]

[117] Probably the Spotted Flycatcher is here referred to, the prefix not being used in a technical sense; it is known here as the Beam-bird, either of which names may be a corruption of the other. Another Norfolk name for this bird is the Wall-bird.

Haue you morinellus marinus or the sea Dotterell better colourd then the other & somewhat lesse [see Note 28].

I send you a draught of 2 small birds the bigger called a Chipper or Betulae Carptor [see Note 48] cropping the first sproutings of the Birch trees & comes early in the spring. The other a very small bird lesse than the certhya or ox eyecreeper called a whinne bird

I send you the draught of a fish taken sometimes in our seas [see Note 69]. pray compare it with Draco minor Johnstoni. this draught was taken from the fish dried & so the prickly finnes less discernible.

There is a very small kind of smelt [see Note 71] butt in shape & smell like the other taken in good plenty about [wh crossed out] Lynne & called Primmes.

Though Scombri Or Makerells [see Note 73] bee a co[=m]on fish yet [in crossed out] our seas afford sometimes strange & large ones as I haue heard from fishermen & others. & this yeare 1668 one was taken at Lestoffe an ell long by measure & presented to a Gentleman a friend of myne.

Musca Tuliparum moschata is a small beelike flye [see Note 108] of an excellent fragrant odour which I haue often found at the bottom of the flowers of Tuleps.

[Fol. 44.] In the little box I send a peece of vesicaria or seminaria marina [yo crossed out] cutt of from a good full one found on the sea shoare [see Note 91].

Wee haue [two or three words smeared out here] also an eiectment of the sea very co[=m]on which is fanago [see Note 91] whereof some very large.

I thank you for communicating the account of Thunder & lightening some strange effects thereof I haue found heere butt this last yeere wee had litle or no Thunder & lightening. [No signature.]

No. V.

DR. BROWNE TO MERRETT.

[This letter which was originally printed in the "Posthumous Works," will be found in MS. Sloane 1911-13, fol. 106, where it is headed in pencil as addressed to Sir Wm. Dugdale, but it was restored to its proper place by

Wilkin in the 1836 Edition of the Works, i., p. 404.]

Honoured Sir

[Fol. 106.] I am sorry I have had [diuersions above] of such necessitie, as to hinder my more sudden salute since I receiued your last. I thank you for the sight of the Sperma Ceti, and such kind of effects from [Lightning & Thunder written above] I have known and about 4 yeares ago about this towne when I with many others saw fire-balls fly & go of when they met with resistance, and one carried away the tiles and boards of a leucomb Window of my owne howse, being higher then the neighbour howses & breaking agaynst it with a report like a good canon. I set downe that occurrence in this citty & country, & haue it somewhere [in crossed out] amongst my papers, and fragments of a woman's hat that was shiuered into pieces of the bignesse of a groat. I haue still by mee a little of the spermaceti of our whale, as also the oyle & balsome wch I made with the oyle & spermaceti. Our whale was worth 500 lib. my Apothecarie got about fiftie pounds in one sale of a quantitie of sperm [see Note 51].

I made enumeration of the excretions of the oake which might bee obserued in england [see Note 112], because I conceived they would bee most obseruable if you set them downe together, not minding whether there were any addition by excrementum fungosum vermiculis scatens I only meant an vsuall excretion, soft & fungous at first & pale & sometimes couered in part with a fresh red growing close vnto the sprouts. first full of maggots in little woodden cells which afterwards turne into little reddish browne or bay flies. of the tubera indica vermiculis scatentia I send you a peece, they are as bigg as good Tennis-balls & ligneous.

The little elegant fucus [see Note 114] may come in as a difference of the abies, being somewhat like it, as also unto the 4 corrallium in Gerard of the sprouts whereof I could never find any sprouts wings Or leaves as in the abies whether fallen of I knowe not, though I call'd it icthyorachius or pisciculi spinam referens yet pray do you call it how you please I send you now the figure of a quercus mar. [inus] or alga which I found by the seashoare differing from the co[=m]on [see Note 114] as being denticulated & in one place there seemes to bee the beginning of some flower pod or seedvessell.

[Fol 106. verso.] A draught of the morinellus marinus or sea doterell I now send you. the bill should not have been so black & the leggs more red, [see Note 28] & [the crossed out] a greater eye of dark red in the feathers of wing and back: it is lesse & differently colourd from the co[=m]on dotterell, wch [wee haue crossed out] cometh to us about March & September. these sea-dotterells are often shot near the sea.

A yarewhelp or barker [some words smeared out] [see Note 39] a marsh-bird the bill 2 inches long the legges about that length the bird of a brown or russet colour.

That which is knowne by the name of a bee-bird [see Note 117] is a litle dark gray bird I hope to get one for you.

That whch I call'd a betulae carptor & should rather have calld it Alni carptor [see Note 48] whereof I sent a rude draught. it feeds upon alder [budds mucaments or written above] seeds which grow plentifully heere & they fly in little flocks.

That [calld by some a written above] whin-bird is a kind of ox eye butt the shining yellow spot on the back of the head [see Note 48] is scarce to bee well imitated by a pensill.

I confess for such litle birds I am much unsatisfied on the names giuen to many by countrymen, and vncertaine what to giue them myself, or to what classes of authors cleerly to reduce them. surely there are many found among us whch are not described; & therefore such whch you cannot well reduce may (if at all) bee set downe after the exacter nomination of small birds as yet of uncertain classe or knowledge.

I present you with a draught of a water-fowl not co[=m]on & none of our fowlers can name it [see p. 79 infra] the bill could not bee exactly expressed by a coale or black chalk, whereby the litle incuruitie [at the end written above] of the upper bill & small recurvitie of the lower is not discerned. the wings are very short, & it is finne footed. the bill is strong & sharp, if you name it not I am uncertaine what to call it pray consider this Anatula or mergulus melanoleucus rostro acuto.

[Fol. 107.] I send you also the heads of mustela or mergus mustelaris mas. et faemina [see Note 21] called a wesel from some resemblance in the head especially of the female wch is brown or russet not black & white like the male. & from their praying quality upon small fish. I have found small eeles small perches & small muscles in their stomacks. Have you a sea phaysant [see Note 22] so co[=m]only calld from resemblance of an hen phaisant in the head & eyes & spotted marks on the wings & back. & wth a small bluish flat bill, tayle longer than other ducks, long winges crossing over the tayle like those of a long winged hawke.

Have you taken notice of a breed of porci solidi pedes.[118] I first obserued them above xx yeares ago & they are still among us. [See also p. 80 infra.]

[118] Mr. Darwin writes ("Anim. and Plants under Domestication," i., p. 78), that from the time of Aristotle to the present day, Solid-hoofed Swine have been occasionally observed in various parts of the world. Dr. Coues also says that this variety seems to be persistent in a Texas breed. See also Professor Struthers in the "Edin. New Phil. Journal," April, 1863. The two distal phalanges of the two great toes, both front and back, in the examples described by Professor Struthers, were joined together, forming a single hoof-bearing bone. The next two phalanges were separate, and sometimes kept widely apart from each other by the introduction of a special ossicle. I have been told that about the year 1827, a breed of solid-footed swine existed at or near Upwell. By some it was thought that their flesh was not good for food because they were "uncloven." Dr. Wren, in a note to Browne's Pseudodoxia (book vi., chap. x.), says, "About Aug., 1625, at a farm 4 miles from Winchester, I beheld with

wonder a great heard of swine, whole-footed, and taller than any other that ever I sawe."

Our nerites or neritae are litle ones [see Note 83].

I queried whether you had dentalia [see Note 83] becaus probably you might haue met with them in england. I neuer found any on our shoare butt one brought mee a few small ones with smooth with [sic] small shells from the shoare. I shall inquire further after them.

Urtica marina minor Johnst. tab. xviii. [see Note 90] haue found more than once by the sea side.

The hobby and the merlin would not bee omitted among hawkes the first coming to us in the spring the other about the autumn. Beside the ospray wee have a larger kind of agle, calld an erne [see Note 3]. I haue had many of them.

Worthy deare Sr, if I can do anything farther wch may bee seruiceable unto you you shall ever readily co[=m]and my endeauours; who am, Sr, Your humble & very respectfull seruant,

THO. BROWNE.

Febr 6 [1668-9.] Norwich.

No. VI.

[MISCELLANEOUS PAPERS. MS. SLOANE 1847, FOL. 198.]

[This volume contains a Miscellaneous collection, mostly letters to his son Edward, and some to "Tom." The following (as all in the volume) is on letter-sized paper, 7-1/2 x 6 in.]

Worthy Sr

[Fol. 198.] Though I writ vnto you last monday. yet hauing omitted some few things wch I thought to have mentioned I am bold to giue you this trouble so soone agayne haue you putt in a sea fish calld a bleak [see Note 74] a fish like an herring often taken with us and eat butt a more lanck & thinne & drye fish.

The wild swanne or elk [see Note 8] would not bee omitted, [here crossed out] being co[=m]on in hard winters & differenced from [the crossed out] our River swanns by the Aspera Arteria. [See also pp. 80 and 83 infra.]

Fulica and cotta Anglorum [see Note 23] are different birds though good resemblance between them, so some doubt may bee made whether it bee to bee made a coote except you set it downe fulica nostras. & cotta Anglorum I pray consider whether that waterbird whose draught I sent in the last box & thought it might bee named Anatula or mergulus melanoleucos may not bee some gallinula. it hath some resemblance with gallina hypoleucos of Johnst Tab 32 [31] butt myne hath shorter wings by much & the bill not so long [Fol. 198 verso] & slender & shorter leggs & lesser & so may ether be calld gallina Aquatica hypoleucos nostras or hypoleucos or melanoleucos Anatula or mergulus nostras.[119]

[119] The "draught" of this bird sent to Merrett is not forthcoming. Professor Newton has been kind enough to send me the following note on this puzzling passage. "Jonston's figure (tab. 31) of Gallina hypoleucos, to which Browne says it bore some resemblance, undoubtedly represents what we know as the Common Sandpiper, Totanus hypoleucus or Actitis hypoleuca, the Fysterlin of the Germans of Jonston's time (p. 160), and Fisterlein or Pfisterlein of modern days. But there seems to be some strange confusion that cannot now be cleared, between this bird and Browne's Anatula or Mergulus melanoleucos [see p. 76 ante], of which some years later, he sent a drawing, under the latter name, to Willughby, in whose work it is described and figured (Lat. Ed. p. 261, Engl. 343, tab. lix.), for this most certainly is the Rotche or Little Auk, Mergulus alle of modern ornithology." In the next letter (p. 81), Browne mentions that he encloses the draft of "Ralla aquatica" here referred to.

Tis much there should be no Icon of Rallus or Ralla Aquatica I haue a draught of one & they are found among us

Feb xii 1668.

The vesicaria I sent is like that you mention [see Note 91] if not the same the co[=m]on fanago resembleth the husk of peas this of [Part crossed out] Barly when the flower is mouldred away. [See also p. 89 infra, where Merrett aptly compares the latter to the flowers of the Grape Hyacinth.]

No. VII.

[BIBLIOTHECA BODLEIANA. MS. RAWLINSON D. cviii. SR THO BROWN TO DR. MERRETT.]

[Fol. 105.] Sr I craue your pardon for this delayed returne unto your last, whose courteus acceptance & worthy entertaynment [?] deserued [a speed blotted out] even a speedier reply. The small plant may fitly come in among the corallines upon the [diff crossed out] account of articulation Icthyorachius [see Note 114] I think will bee a good Diference [?]. whether you will subexpand [?] the word I referre it to yourself. certhia may best bee vertice aureo [word blotted out] or vertice aureo penicello vix imitando. morinellus marinus [see Note 28] I think rather then Aquaticus becuse it is seen most about the sea coast. Anas alis oculatis[120] rather then Anser for it is not altogether so longe as a wild duck. of porci solidipedes [see Note 118] there are still in this country in some places. and I am promised a pigge by a Gentleman that hath still a boar and sow of that kind. I tooke notice of them 26 years ago & having not lately [met with crossed out] met with any thought the race had been worne out butt I perceue it is not--they are whole footed in the forfeet & have [only crossed out] a seame only in the hinder. so they are animalia duplici nomine i[=m]unda. The wild swans or elk [see Note 8] in [very crossed out] lasting cold winters are most plentifull. It is larger then the River swan somewhat gray & of a lowder note & [differenced call crossed out] a recuruation of the Aspera arteria in the sternon as I noted in the margin long agoe in vulgar errors. the blicca marina [see Note 74] may well be named Harengiformis. [several words smeared out] I have the draught of that an Herring & a pilcher in one paper upon that account [Fol. 104 verso] I belieue [?] you were well informd of the cotta [see p. 79] & fulica of our Ralla Aquatica I enclose a draught.

[120] Possibly the Pintail, Dafila acuta (Linn.), see p. 77.

Of porci solidipedes there are diuers still in the country in some places I am promised a pigge by a friend who cherisheth that [new crossed out] breed. I tooke notice of them 26 yeares ago, & hauing not lately minded them thought they had been worn out butt I perceiue they are not--some are more plainly wholefooted then others & especially in the fore feet & in the rest there is no thorough fissure butt at most a superficiall seame, so they are [No. 3 cap 27 above] Quadrupedia duplici nomine i[=m]unda.

[This last paragraph seems to have been written by way of emendation of what appears above on the same subject. A photograph of a portion of the above letter will, by the courtesy of the Bodleian Librarian, be found as a frontispiece to this volume. Mr. Jenkinson, the Librarian of the University of Cambridge, and through him, Mr. G. F. Warner and Mr. Kenyon, of the Department of Manuscripts of the British Museum, have kindly interested themselves in the transcript of this letter, which was very difficult to decipher.]

No. VIII.

BIBLIOTHECA BODLEIANA (MS. RAWL. D. cviii.)

[Draft of a letter from Sir Thomas Browne, described in the Catalogue of the Rawlinson MSS. as to the Secretary of the Royal Society, but from its contents evidently written to Merrett, whose letter, dated 8th May, 1669, is in part a reply to it.]

[Fol 58.] Honord Sr I humbly thank you for your care of my sonnes paper & the Royll Societie for their acceptance of it. If hee bee in health I knowe hee is mindfull of their co[=m]ands receiued aboue 2 months ago by a letter from Mr. Oldenburg.[121] I haue not heard from him of late the last I receiued was from Komorn[R] in Lower Hungary and hee was then going to the mine countryes. I think the Rowd may bee calld Rutilus ventre magis compresso[122] wch is the first discoverable difference to the eye. The weazelling [see Note 60] is as you see in the draught a long fish figura ad teretem vergente. somewhat of the shape butt differing in the head from the mustela viuipara of Schoneueld. butt not lozenged on the back though the back bee much darker then the other parts. I send you the figure of the head of a cristated wild duck. it is black blackish [sic] in the greater part of the body some white on the brest & wings blewish legges & bill & seems to bee of the Latirostrous tribe perhaps you haue it not. it may bee called Anas macrolophos [Fol. 59] as excelling in that kind.[123] there is also a draught of one sort of mergus cristatus resembling that of Aldrovandus or Johnstonus where there is only the figure of the head only this is also ruffus butt the head sad red.[124] wee haue a kind of teale which some fowlers call crackling teale from the noyse it maketh[125] it is almost of the bignesse of a duck coming late of the yeare & latest going away hath a russet head & neck with a dark yellow stroak about a quarter of an inch broad from the crowne to the bill winged like a teale a white streake through the middle of the wings and edges thereof the tale blackish. it may be calld Querquedula maior serotina. I send you the figure in litle of a pristis[126] wch I receaued from a yarmouth seaman. you may please to compare it wth yours. the asper you mention is much like our Rough or Aspredo.

[121] Henry Oldenburg (1615-1677) was born at Bremen. Came to England about 1640, where he remained eight years. In 1653 he was sent to England from Bremen on a diplomatic mission to Cromwell. He returned to England a third time in 1660. He was an original Member of the Royal Society, and became one of its first Secretaries. A half-length portrait is in the possession of the Royal Society.

[R] A well-known town on the Danube, forty-seven miles west of Buda-Pesth, probably the Comorra of E. Browne's letter to his father, cf. Wilkin, i., p. 159.

[122] The Rudd (Leuciscus erythrophthalmus, Will.) is known in Norfolk as the Roud. Browne seems to treat it as a variety of the Roach (Rutilus, Willugh.), and Merrett in his second letter remarks with approval "you have very well named the Rutilus."

[123] Fuligula cristata (Linnaeus), the Tufted Duck.

[124] Professor Newton suggests that Browne intended to write Mergus cirratus. Aldrovandus figures the head, iii., p. 283, and that of M. longirostris in the preceding page. This last is copied by Jonston (fol. 47). Both birds seem to be female or immature Goosanders. Neither author has a M. cristatus.

[125] The above description certainly applies to the Common Teal, which was well-known to Browne (vide supra, p. 14), and that species is with us all the year; I cannot help thinking, however, that he had in his mind the Garganey, or Summer Teal, so called from the season of its visit to us. This species is known to the Norfolk gunners as the "Cricket Teal," and being slightly larger than the common species it might well be called by him "Querquedula major serotina."

[126] See Note 55, p. 36. It will be noticed that both this and the Centriscus mentioned at p. 41 were given to Browne by a "seaman of these seas," but may possibly have been brought home as curiosities from a foreign voyage; the Saw-fish, however, mentioned at p. 36, is distinctly stated to have been "taken about Lynn." It is a matter of intense regret that the numerous drawings mentioned in these letters should have been lost.

I forgot in my last to signifie that an oter [an other?] Elk or wild swan was headed like a goose that is without any knobb at the bottome of the bill. [See p. 80 and Note 8.]

Haue you had the duck called Clangula in Ald. [drovandus] & Johnst.[127]

wee haue one heere wch answereth their descriptions exactly butt [i.e., except] only in the colour of their leggs & feet.

[127] Aldrovandus's figure of "Clangula" (head only, iii., p. 224) is too indefinite for determination. He says the feet are yellow, but Jonston, who refers to it under the name of Anas platyrhincus describes it fairly well (p. 145). Clangula ab alarum clangore, Aldrov., i.e., "Rattlewings," an old name by which the Golden-eye was known to the Norfolk gunners.

Haue you a willock a sea fowl like a rook or crowe.[128]

[128] A local name for the Guillemot. Merrett says, in a letter dated 8th May, 1669, "The Clangula I know no more of than reading hath informed mee; [see Note 127] a willock I have seen brought from Greenland,[S] where they are said exceedingly to abound, but never thought either of them was found in England, and having not taken sufficient notice of the latter, crave your description of both."

[S] The Greenland of those days was Spitsbergen, where they would be met with by the Whalers, but in that case the bird would be Bruennich's Guillemot, a species not then differentiated.

No. IX.

[MISCELLANEOUS PAPERS. MS. SLOANE 1847, FOL. 182.]

[Fol. 182.] Sr I craue your pardon that I haue no sooner sent unto you. I shall be very reddie to do you service in order to your desires And shall endeavour to procure you such animalls as I haue formerly met with & any other not ordinary wch [shall crossed out] are to bee acquired. though many of my old assistants are dead. & sometimes they fell upon animalls, [not to bee crossed out] scarce to bee met with agayne. I wish I had been acquainted with your desires 3 yeares ago. for I had about fortie hanging up in my howse. wch the plague being at the next doores the person intrusted in my howse, burnt or threw away. The figure of the weasell Cray [see Note 60 and p. 82] was in a long paper pasted together at the ends & I make no question you will find it otherwise I would send another [the willick wee in crossed out] that fowl wch some call willick, [see Note 128] wee meet with sometimes. The last I met with was taken on the sea shoare. the head and body black the brest inclining to black headed and billd like a crowe, leggs set very backward wings short leggs set very backward (sic) that it move overland very badly only. it may bee a kind of cornix marina. [The latter portion very badly written and difficult to decipher.]

[Fol. 184 verso.] That litle plant upon oyster shells [see Note 91] I remember I haue seen & surely is some kind of vescaria or calicularia

of what that other [was crossed out] electricall body was Mr. Boyle[129] showed [smear] by this time more tryall hath probably been made, something of jet it might consist of.

[129] The Hon. Robert Boyle (1627-1691), although deeply learned in many branches of science, was chiefly distinguished as a chemist. He took a leading part in the founding of the Royal Society, and was elected President in 1680, but from some conscientious scruple did not accept the office. Naturalists are deeply indebted to him, as he was "the first that made trial of preserving animals" in spirit (see Grew's "Musaeum Regalis Societatis" (London, 1681), p. 58).

I thank you that you were pleased to enquire of those German gentlemen concerning my sonne I receiued a letter lately from him he hath not been unmindfull of the R. Society's co[=m]ds & hath been in Hungaria in the mines of Gold, sylver & copper at Schemets, Cremitz & Neusol & desired mee to signifie so much to Mr. Oldenberg.

[The above is hastily scrawled; it was evidently indited to Merrett, as indicated by the reference to the German gentlemen, &c.; the date would therefore be some time in the year 1669. Wilkin prints it in the 1836 Edition,

Vol. i., p. 408, but it is not in Bohn's reprint.]

APPENDIX A.

[TWO LETTERS FROM DOCTOR CHRISTOPHER MERRETT TO SIR THOMAS BROWNE, MS. SLOANE 1830, FOL. 1 TO 3. THEY ARE BOUND UP IN INVERSE ORDER OF DATE.]

[Reply to No. 2 in the above Series.]

[Fol. 3.] WORTHY SR,--y^{rs} of ye 14th instant I recaeved as full off learning in discovering so many very great curiosities as kindness in communicating them to mee & promising y^r farther assistance. ffor which I shall always proclame by my tongue as well as by my pen, my due resentment & thanks.

The 2 funguses [guses crossed out and i inserted] y^w sent ye figures off [see Note 106] are ye finest & rarest as to their figure I have ever seen or read of, & soe is y^r fibula marina, far surpassing one I reacived from Cornwall much of ye same bigness, neither of which I find anywhere mentioned. The urtica marina minor Jonst. & physalus I never met with, nor have bin informed off y^e canis charcharius alius Jonst. Many of y^e Lupus piscis I have seen, & have bin informed by y^e Kings fish monger they are taken on our coast, but was not satisfyed for some reasons off his relation soe as to enter it into my pinax, though tis said to bee peculiar to y^e river Albis [= Elbe] yet I thought they might come sometimes thence to y^r coasts. Trutta marina I haue and y^e loligo, sepia, & polypus y^e 3 sorts off y^e molles have bin found on our western coasts which shall bee exactly distinguished--As for y^e Salmons taken a bove London towards Richmond & nearer, & y^t in great quantity some years they have all off them their lower jaw as y^w observ, [see Note 92] & our fishermen [men crossed out] say they usually wear off some part off it on ye banks or els ye lower would grow into ye upper & soe starve them as they have sometimes seen--y^w ask whether I haue ye mullus ruber asper, or ye piscis Octangularis Wormii. or ye sea worm longer than ye earth worms, or ye garrulus Argentor. or ye duck cal'd a May chit or ye Dor hawke. The 4 first I haue noe account off ye 2 later I know not especially by those names, wee have noe hawk by y^t name [see Note 42] y^r account of succinum as all ye rest will bee registered. As for ye Aquila Gesneri I never saw nor heard off any such in ye Collidge for [fol. 3 verso] this 25 years last past. Sr y^w are pleased to say y^w shall write more if y^w know how not to bee surpurfluous--certainly what y^w have hitherto done hath bin all curiosities, & I doubt not but y^w have many more by you--I can direct y^w noe further than y^r own reason dictates to y^w. Besides those mentioned in ye pinax I have 100 to add, & cannot give y^w a particular off them--whatever y^w write is either confirmative or additional. I doe entreat this favour off y^w to inform mee fuller off those unknown things mentioned herein, & to add ye name page &c of ye Author if mentioned by any or else to give them such a latin name for them as y^w have done by ye fungi which may bee descriptive & differencing off them. Sr I hope ye publigs [sic] interest & y^r own good genius will plead y^r pardon desired by

y^r humble servant

CHR. MERRETT.

London Aug. 29. 68.

[Reply to No. 8 of the above Series.]

[Fol. 1.] WORTHY SR,--my due thanks premised I at present acquaint y^w y^t y^w have very well named ye Rutilus & expressed fully ye cours to bee taken in ye imposition of names viz ye most obvious & most peculiar difference to ye ey or any other sens. I am farther to say y^t ye icon of ye weazeling came not to my hands, pray bee pleas'd to look amongst y^r papers perhaps it might bee laid by through some accident or other [I have added above] ye figures of y^r anas macrolophos, & of ye mergi cristati [see Note 124] & of ye pristis y^t which came from Cornwall was of ye gladius, ye name of sword fish being applied to both of them by our nation. It seemeth by y^w y^t ye Norwich aspredo is not ye Ceruna fluviatilis contrary to what Camden affirms, for ye rutilus mentioned in mine to y^w differs toto coelo from ye ceruna--The difference of ye Elks bill by y^w signified is remarkable to distinguish it from others of its own kind. [See p. 83 supra.] The crackling teal seems [clearly crossed out] to bee ye same which Dr Charleton[130] mentions in his Onomasticon under ye name of ye cracker,& showing him y^r description hee acknowledged to bee ye same, ye clangula I know noe more of than reading hath informed mee, a willock I have seen brought from Greenland where they are said exceedingly to abound, but never y^t [thought?] either of them was found in England, & having [not added above] taken sufficient notice of it ye later, crave y^r description off both.

[130] In Charleton's "Onomasticon," at p. 99, the Cracker is called by him, Anas caudacuta, and is said to be the "Gaddel" of the London dealers in fowl. [See Note 125.]

And now Sr since my last only 2 things remarkable haue come to my knowledge. The one was a cake off black amber 1/6 off an inch thick & neer a palm each way. Mr. Boyle brought it to ye R. society to whom it was sent from ye Sussex shore, hee had only tryed it to its electricity & found it answer his expectation, farther tryals will be made of it. The second is a small plant found on oyster shells which when fresh did perfectly represent ye flowers off Hyacinthus botryoides, [see Note 91] but y^t was somewhat longer & not so much sweld out towards its pedunculus, some of them are here inclosed. Tis doubtless a sort off vesicaria, though much different from what y^w sent mee. Most off them are now shrunk & ye sides constituting ye cavity come together & appear only a transparent husk. One thing more I had to add (but scarcely dare speak it out) y^t is if it would please [you added above] to let it bee done without y^r charge & 2ly if it might be done without y^r trouble, then I would beg off y^w to set some a work to procure mee some of those rare animals &c y^w have mentioned in your seueral Letters. My intention therein is double: first to take their descriptions & furnish our colledge with them as curiosities, all being lost by ye fire this is onely wished but must not bee proposed without ye former limitation by y^r too much allready obliged friend & servant

8th May '69.

CHR. MERRETT.

I met this week with some persons off quality high Germans who lately saw y^r son & record all good things off him.

ffor Dr Browne off Norwich.

[The reply to this letter is No. IX of the above Series.]

APPENDIX B.

[MISCELLANEOUS PAPERS. MS. SLOANE 1847, FOL. 56-57.]

[See Note 51, p. 32 supra.]

Praye Request Mr. Johnson to obtayne this fauor of Mr. Bacon who is unknown to mee, to afford mee his resolution to these few queries concerning the whale [wch crossed out] whereof I understand he had the cutting up and disposure whether there were any spermacetie found, or made out of other parts beside the head; if soe, of what parts & out of what most: and whether any out of the meere fleshie parts whether that wch runne from it about the shoare came out of the mouth.

[Not signed or dated.]

REPLY.

Sr in Answer to your questions conserninge the whale, I founde noe Sper[=m]e but in his heade and that after I had taken off his scalpe one tonn weight [or more written above] of a nexuous substance, we found in the circumference as large as a small coach wheele in the middle part certain round pieces of Sperm as bigge as a mans fist some as large as eggs and on the out side of the said rounds, flakes as large as a mans head in forme like hony combs being very white and full of oyle. And that Sp. wch was cast upon the shore I doe conceive came out of his nostrells. thus much ffrom him who doth remayne Sir your humble Servant, Arthur Bacon Yarmouth 10th May 1652.

BROWNE TO DUGDALE ON CERTAIN FOSSIL BONES.

["EASTERN COUNTIES COLLECTANEA," pp. 193-195].

The letter referred to in the foot-note on page 33, written by Sir Thomas Browne to Dugdale, and formerly in the possession of the late Mr. Arthur Preston of Norwich, whose collection of manuscripts was dispersed by auction in August, 1888, was printed in a brief-lived and little-known local publication, entitled the "Eastern Counties Collectanea" (1872-3), at page 193. In this letter occurs a passage which confirms the doubt expressed as to the Whales which had young ones after coming on shore at Hunstanton being Sperm Whales. They are expressly said to have been of that sort "which seamen call a Grampus," and as Sir Nicholas le Strange, in a MS. preserved in the Muniment room at Hunstanton, applies the name "Grampus" to an undoubted specimen of Hyperoodon rostratus (as shown both by his description and outline sketch) which came ashore there in the year 1700, I have little doubt that the Cetaceans in question belonged to that species and not to Physeter macrocephalus.

This letter is interesting also as filling a gap in Wilkin's series and I therefore reproduce it, omitting only occasional learned digressions which do not affect the subject. The original not being available, I have used the copy in the "Collectanea" before mentioned.

Dugdale, in November, 1658, and again later, had written to Browne, sending him a bone of a "fish which was taken up by Sir Robert Cotton, in digging a pond at the skirt of Conington downe," and asking his opinion thereof. (Wilkin, i., pp. 385 and 390.)

To the first of these letters Browne replied, under date of the 6th December, 1658, "I receaued the bone of the fish, and shall giue you some account of it when I have compared it with another bone which is not by mee" (op. cit. p. 387). The letter which follows and which was unknown to Wilkin supplies this information.

[p. 193.] "Sr I cannot sufficiently admire the ingenious industry of Sr Robert Cotton in preserving so many things of rarity and observation nor commend your own enquiries for the satisfaction of such particulars. The petrified bone you sent me, which with divers others was found underground, near Cunnington, seems to be the vertebra, spondyle or rackbone of some large fish, and no terrestrious animal as some upon sight conceived, as either of Camel, rhinoceros, or elephant, for it is not perforated and hollow but solid according to the spine of fishes in whom the spinal marrow runs in a channel above these solid racks, or spondiles.

"It seems much too big for the largest Dolphins, porpoises, or sword fishes, and too little for a true or grown whale, but may be the bone of some big cetaceous animal, as particularly of that which seamen call a Grampus; a kind of small whale, whereof some come short, some exceed twenty foot. And not only whales but Grampusses have been taken in this Estuarie or mouth of the fenland rivers. And about twenty years ago four were run ashore near Hunstanton and two had young ones after they came to land. But whether this fish were of the longitude of twenty foot (as is conceived) some doubt may be made for this bone containeth little more than an inch in thickness, and not three inches in breadth so that it might have a greater number thereof than is easily allowable to make out that longitude. For of the whale which was cast upon our coast about six years ago a vertebra or rackbone still preserved, containeth a foot in breadth and nine inches in depth, yet the whale with all advantages but sixty-two foot in length. [p, 194.] We are not ready to believe that, wherever such relics of fish or sea animals are found, the sea hath had its course. And Goropius Becanus[131] long ago could not digest that conceit when he found great numbers of shells upon the highest Alps. For many may be brought unto places where they were not first found.

[131] This seems to refer to the "De Gigantibus eorumque reliquiis" of J. van Gorp, Jean Becan, or Joannes Goropius (as the name is variously given in the "Biographie Universelle" (b. 1518, d. 1572), and apparently published after the Author's death by Jean Chassanion, 8vo, Basileae, 1580, and another edition in 1587. See Brit. Mus. Cat.; but I have not seen the book.

"Some bones of our whale were left in several fields which when the earth hath obscured them, may deceive some hereafter, that the sea hath come so high. In northern nations where men live in houses of fishbones and in the land of the Icthiophagi near the Red sea where mortars were made of the backbones of whales, doors of their jaws, and arches of their ribs, when time hath covered them they might confound after discoverers....

"For many years great doubt was made concerning those large bones found in some parts of England, and named Giants' bones till men [p. 195] considered they might be the bones of elephants brought into this island by Claudius, and perhaps also by some succeeding emperors [then follow other ancient examples of the finding 'elephants bones' in various countries attributed to similar modes of introduction]. But many things prove obscure in subterraneous discovery....

"In some chalk pits about Norwich many stag's horns are found of large beams and branches, the solid parts converted into a chalky and fragile substance, the pithy part sometimes hollow and full of brittle earth and clay. In a churchyard of this city an oaken billet was found in a coffin. About five years ago an humourous man of this country after his death and according to his own desire was wrap't up in a horned hide of an ox and so buried.[T] Now when the memory hereof is past how this may hereafter confound the discoverers and what connjectures will arise thereof it is not easy to conjecture.

[T] Richard Ferrer, of Thurne, by his will, proved about 1654, directed that his "dead body be handsomely trussed up in a black bullock's hide, and be decently buried in the Churchyard of Thurne."--"Norfolk Archaeology," v., p. 212.

Sr Your servant to my power,

THO. BROWNE."

This is endorsed "Sr Thomas Browne's discourse about the Fish bone found at Conington Com. Hunt, Shown, Dr. Tanner."

APPENDIX C.

[SLOANE MS. ADDITIONAL 5233, LARGE FOLIO, IS A VOLUME LABELLED "DR. EDW. BROWN'S DRAWINGS."]

"Some original drawing of Towns, Castles, Antiquities, Medals &c. by Dr. Edward Browne in his Travels & presented by his Father Sir Thomas Browne. Who hath write upon sev^{ll} of them what they are."

The above is the inscription written on the fly-leaf of this volume, which I hoped might have contained some drawings of birds or fishes by Sir Thomas Browne, but there is nothing in it of interest from a Natural History point of view. In Wilkin's Catalogue of the MSS. (Vol. iv., p. 476) it is described as "a collection of very curious drawings (some coloured) of public buildings, habits, fishes, mines, rocks, tombs, and other antiquities, observed by Sir Thos. and Dr. Edward Browne in their travels," but there are no fishes, birds, or other animals in the volume.

APPENDIX D.

Draft of a letter from Sir Thomas Browne to his daughter Elizabeth, enclosing two pictures of a Stork. This and the next letter are in the Bodleian Library (MS. Rawl. D. cviii.)

[Fol. 70.] This is a picture of the stork [see Note 14] I mentiond in my last. butt it is different from the co[=m]on stork by red lead colourd leggs and bill[132] and the feet hath not vsuall sharp poynted clawes butt resembling a mans nayle, such as Herodotus discribeth the white Ibis of AEgypt to haue. The ends of the wings are black & when shee doth not spred them they make all the lower part of the back looke black, butt the fethers on the back vnder them are white as also the tayle. it fed upon snayles & froggs butt a toad being offered it would not touch it. the tongue is about half an inch long. the quills of the wing are as bigge or bigger then a swans quills. it was shott by the seaside & the wing broake. Some there were who tooke it for an euell omen saying If storks come ouer into England, god send that a co[=m]onwealth doth not come after.[U]

[132] Browne evidently was not very familiar with the Stork, which is not surprising, seeing that it is a very rare bird in Britain; it may be that he had only seen the bird in its immature stage, for the "red-lead" hue of the legs is very characteristic of the adult bird. [See also Note 14, p. 10.]

[U] In reference to the Dutch fable of those days that Storks would only inhabit republican countries.

That picture with the lesser head is the better.

MS. RAWL. D. cviii.

Draft of a letter containing further particulars with regard to the Stork. There is nothing to indicate to whom it was addressed.

[Fol. 77.] A kind of stork was shott in the wing by the sea neere Hasburrowe & brought aliue vnto mee. it was about a yard high red lead coloard leggs and bill. the clawes resembling human nayles such as Herodotus describeth in the white Ibis of AEgypt The lower parts of the wings are black which gathered up makes the lower part of back looke black butt the tayle vnder them is white as the other part of the body. it fed readily upon snayles & froggs, butt a toad being offered it would not touch it: the tongue very short [not crossed out] an inch long. it makes a clattering noyse by flapping one bill agaynst the other somewhat like the platea or shouelard.[V] the quills [about crossed out] of the biggnesse of swans bills [sic quills?] when it swallowed a frogge it was sent downe into the stomak by the back side of the neck as was perceaued upon swallowing. I could not butt take notice of the conceitt of some who looked upon it as an ill omen saying if storks come ouer into England, pray god a co[=m]on wealth do not come after.

[V] The Spoonbill.

In addition to these letters there are in the Bodleian Library a letter from Elizabeth Browne to her brother, describing the above-mentioned Stork, and desiring him to keep one of the two pictures himself, and to give the other to his sister Fairfax (MS. Rawl. D. 108, fol. 71), and a draft of a letter from Sir Thomas Browne about a remarkable fly (see ante p. 68 and Note 110), which offended the cattle extraordinarily, found at Horsey Marshes (MS. Rawl. D. 108, fol. 103). There is also (MS. Rawl. D. 391, fol. 55) a letter from Sir Hamon le Strange to Sir T. B., dated Jan. 16, 1653. About half this letter is printed by Wilkin, i., pp. 369-70. He mentions towards the end that he sends certain observations on T. B.'s "Enquiries into Common Errors," at page "27 thereof I write of a whale cast upon my shoare." This criticism is now separated from the letter, which originally covered it, but happily is preserved in the British Museum, MS. Sloane, 1839. fols. 104-145.

www.ingramcontent.com/pod-product-compliance
Lightning Source LLC
Chambersburg PA
CBHW070324190526
45169CB00005B/1734